T0115865

UNDER A GREEN SKY

UNDER A

GREEN SKY

GLOBAL WARMING, THE MASS EXTINCTIONS OF THE PAST,
AND WHAT THEY CAN TELL US ABOUT OUR FUTURE

Peter D. Ward

HARPER ● PERENNIAL

NEW YORK ● LONDON ● TORONTO ● SYDNEY ● NEW DELHI ● AUCKLAND

HarperCollins books may be purchased for educational, business, or sales promotional use. For information please email the Special Markets Department at SPsales@harpercollins.com.

First paperback edition published 2008. Reprinted in Harper Perennial 2017.

Designed by Daniel Lagin

The Library of Congress has catalogued the hardcover edition as follows:

Ward, Peter Douglas, 1949.
   Under a green sky : global warming, the mass extinctions of the past, and what they can tell us about our future / Peter D. Ward.—1st Smithsonian Books ed.
      p. cm.
   Includes bibliographical references and index.
   Contents: Welcome to the revolution!—The overlooked extinctions—The Mother of all extinctions—The misinterpreted extinction—A new paradigm for mass extinctions—The driver of extinctions—Bridging deep past with near past—The oncoming extinctions of winter—Back to the Eocene.
   ISBN: 978-0-06-113791-4
   1. Extinction (Biology) 2. Paleoclimatology. 3. Global warming—Environmental aspects. 1. Title.

QE721.2.E97W384 2007
577.276—DC22

2006052250

ISBN 978-0-06-113792-1 (pbk.)

HB 09.12.2022

Contents

Contents

The climate is like a wild beast, and we're poking it with sticks.

—CLIMATOLOGIST WALLY BROECKER

INTRODUCTION

Going to Nevada

Seattle-Tacoma International Airport at dawn: Even in the gloaming, a bright, sterile, exceptionally hideous example of twenty-first-century American architectural ugliness seems like a suitable send-off point toward a past perhaps even grimmer than our climatic present—but perhaps no more so than our possible future. The waiting lines, the ritual undressing of shoes and belt, the blank scrutiny of identification and tickets, followed by the cattle-like entry into the flying silver tube to find the assigned middle seat between well-stuffed strangers for the supposedly short flight, giant engines snarling out clear but heat-soaking vapors of jet engine residue into the atmosphere, and from the window now high above the world, an amazing sight to a Pacific Northwest native: the high Cascades virtually without snow on this early April 2005 day, following the warmest and driest winter in Pacific Northwest history, ski areas going broke as rock skiing loses clientele not pleased with the necessary artificial ice at Snoqualmie, Stevens Pass, Whistler Blackcomb, Grouse Mountain, and Crystal Mountain ski areas, among many others showing summer rocks in winter. Even Mt. Rainier seems rockier than usual, its glaciers beating a hasty retreat and leaving behind 12,000 years of rocks, airplanes, and human or other animal frozen food long ago lost. The whole, dry mountain range, visible to our Nevada arrival, flaunts its uncovered geology until we circle the Reno basin, touch down; the slow exit from the plane into a different ugliness where the volume and brightness of the movie we have found ourselves in has been jacked off scale in fine William Gibson style. Reno-Tahoe International Airport, where even the gates are stuffed with slot machines screeching a cacophony of enticement at frantic decibel overkill, electricity be damned. Out of the airport to the rental car, a huge sport-utility vehicle, of course, and for once a necessity for where we are going.

We rocket out of the parking lot, screaming through Reno on the freeway east, passing quickly into the empty rat lands of the sorely missed Hunter S. Thompson, tripping out at the absolute ugliness of a landscape repellant to begin with that has had twisted, rusting metal hulks of unknown ancestry sprinkled among the itinerant whorehouses and casinos in a random pattern across its waterless salt flats and outcrops. Two hours of driving fast (but not fast enough, as muscle cars snarling their high-speed anxiety whiz past toward nowhere and everywhere) brings us to Hawthorne, Nevada, whose largest structure is of course the casino, cigarette smoke venting from its few stained windows like some belching coal-fired Oliver Twist factory plant of Dickensian England, past the one museum in town—slower now, ogling the Armament Museum, where at least one model of every shell casing ever used by the town's biggest employer, the U.S. military, sits in forlorn splendor all with flowers bravely growing from the brass openings on top, a '60s dream come true. To the biggest motel in town to toss now-opened bags onto swaybacked beds, liberating the boots, leather, vests, and cold steel anathema to airline carry-ons, and now looking like Halloween imitations of desert rat miners, we point the car east and south, and for mile after tens of miles pass the damnedest-looking B-movie bunkers extending as far as the eye can see, seemingly millions of the squat concrete burial mounds marking the storage of unknown tons of live munitions in quantity probably second only to that held by the insurgents in Iraq. An hour of this, finally into Luning, and damn, the Luning Bar, looking like it always has (early and late Nevada spider-webbed rattrap décor), sits closed, so no eye-opener on the way to the outcrop.

We leave the highway and all those muscle recreational vehicles around us that are making the long trek to Vegas across the Nevada no-man's-land and shoot onto the wide, pale playa, an old lake bed of Ice Age antiquity that stands between us and the hills ahead, the raucous backseat crew calling without success for a few 60-mile-per-hour wheelies in the lake bed. The track becomes fainter, and we enter the hills in four-wheel drive, the motor growling in protest as we lurch into high canyons, while the navigator beside me is covered with maps and barking directions over the din, impatience thick now, to a turnout well known from past trips here, the setting-out point for the trail to Muller Canyon, the best example of rocks clutching one of the five largest of all mass extinctions, that at the end of the Triassic period, a catastrophe of 200 million years ago conveniently blamed on a Big Rock From Space smashing into a Triassic world populated by early mammals and dinosaurs as well as croc-like beasts galore on land and oceans of ichthyosaurs, ammonites, and strange flat clams, secure as such dumb brutes can be, not knowing that their world was one day from over according to twenty-first-century cant, the only problem being that our previous trips to this barren place did not yield the faintest whiff of iridium or glassy spherules or shocked quartz or impact layers so visible in that other known impact extinction, that at the end of the Cretaceous.

Sidestepping across the high hills on the faint path through the piles of strata all around, rocky layers once neatly and horizontally ordered but now layers akimbo, fractured with faults, and burrowed with Saddam spiderholes made by Cowboy Age miners looking for riches in the worst possible place to find material wealth and the best possible place to disinter the dead and interrogate them about the identity of their killer. Mountain sheep jump in fright as we come over the last hill onto the steep slope of our target outcrop—damn and finally—hundreds of feet of limestone sandwiching a 60-foot-thick band of mudstone containing some level where the Triassic ends and the Jurassic begins, and the realization yet again that this is another of the planet's stony cemeteries. A long scorpion pit where we dug in search of this supposed disaster level last time here, a trench now permanently part of the landscape, but in the sins committed against our planet, it hardly registers. The limestones above and the limestones below are packed with life, mainly mollusks, a good Triassic fauna below, a good Jurassic fauna above, and what a supreme difference those two worlds show with clearly almost no survivors of some catastrophe grabbing the river of life and giving it a 90-degree kink into a whole new assemblage of life, the real start to the Age of Dinosaurs after the experimental mucking about in drifting evolution that was the Triassic.

So how about that 60-foot-thick siltstone, almost bereft of fossils—what caused it? But a year or so ago the answer would have been knee-jerk recital: The fossilized dead bodies are evidence of a mass extinction, and since the groundbreaking 1980 discovery of the Alvarez team from Berkeley that the Age of Dinosaurs was ended by an asteroid strike from space, the geological fraternity has pronounced all mass extinctions to have been guilty of asteroid impacts until proven otherwise. Now we are not so sure, for none of the telltale clues of such a cosmic event are in evidence here. Yet if not asteroids or comets from space, what? Exonerating the asteroids leaves but a few suspects, and by 2005 one deemed most likely was, and remains, rapid climate change and *really* fast global warming brought about by not a little methane and a whole lot of carbon dioxide poured into this world from volcanic smokestacks and deep sea bubbles of poisonous greenhouse gas burped out of the sea but one of a series of such mass extinctions, greenhouse extinctions, the rule not the exception, and a road we humans might again travel on now, seemingly oblivious to the road washout ahead, an accident about to happen one more time, or, if we interpret the rock record correctly, many more times.

From the top of our outcrop a valley spreads out, and in the distance the ribbon of road we had left still carries the endless number of cars toward Vegas and the chance to roll the dice, to hit the jackpot, but some number of them will bust instead, just as the Triassic world did, a bust that meant the death of 60 percent of all species on Earth. And guess what—our world is rolling the same set of dice.

In this book I will marshal the history of discovery, beginning in the 1970s, that has led an increasing number of scientists across of broad swath of fields to conclude that the past might be our best key to predicting the future. As strewn across this barren, nearly lifeless hillside in the nontouristy middle of Nevada, if there is even the slightest chance that the carbon dioxide in Earth's atmosphere of 200 million years ago caused this mass extinction, as well as numerous other times before and since that ancient calamity, then it is time for we practitioners who study the deep past to begin screaming like the sane madman played by Peter Finch in the classic 1976 film *Network*, who brought forth his pain with the cry: "I'm as mad as hell, and I'm not going to take this anymore."

In our case, this cry must be: "I am scared as hell, and I am not going to be silent anymore!"

This book is my scream, for here in Nevada, on that day when heat was its usual quotidian force of death, we sat on the remains of a greenhouse extinction, and it was not pretty, this graveyard, the evidence clutched in these dirty rocks utterly demolishing any possibility of hyperbole. Is it happening again? Most of us think so, but there are still so few of us who visit the deep past and compare it to the present and future. Thus this book, words tumbling out powered by rage and sorrow but mostly fear, not for us but for our children—and theirs.

CHAPTER 1

Welcome to the Revolution!

ZUMAYA, SPAIN, JULY 1982

A warm but wet wind from the sea, a wind pushing gray scudding clouds onshore from the squall-torn Bay of Biscay greeted the two geologists as they slowly drove through the narrow, building-lined streets of a small, tiled Basque town named Zumaya, in the quiet of an early Sunday morning. Their knees were still cramped from the daylong drive of the day before, when they had crossed the neck of France by a route that began on the sun-kissed Mediterranean coast at Banyuls-sur-Mer in the Languedoc region, then clung to the edges of the rugged Pyrénées Mountains for their entire south to north length before ending late that night at a cavernous and gloomy hotel perched on the stormy Atlantic Ocean coast in the Basque city of San Sebastián, Spain.

One of the two was Jost Wiedmann, a famous German paleontologist from Tübingen University, itself the most famous and storied paleontological center in the world. He had spent his career studying the geological ranges of one particular group of fossils, one of the most celebrated of all fossil groups, the ammonite cephalopods. He practiced the standard methodology of his German predecessors: studying the collection of the fossils from known locations in strata to produce a "biostratigraphy," literally the differentiation of the many great piles of sedimentary or layered rock so prodigiously scattered across Earth's crust. His particular interest was mass extinction, those short-term biotic catastrophes that were the most dramatic bookmarks in the tables of strata. He had spent much of his fieldwork among the strata of the Cretaceous period making up the fabulously beautiful coastline of France and Spain known as the Basque Country, a place inhabited by a dour race still wishing to be known as a country separate from either France or Spain.

Wiedmann had published widely reports that the ammonites showed no evidence of a rapid extinction but of something quite different. In a number of famous papers that had been published in journals read not just by the small band of professional paleontologists but also by a far wider spectrum of geologists and evolutionary biologists, Wiedmann had presented evidence that the final extinction of the ammonites was the final act of a long, slow diminution of diversity that had lasted more than 20 million years. By the end, almost none were left anyway, making the K-T event (an event straddling the Cretaceous and Tertiary periods) a minor extinction at best—at least for the ammonites.

I was the other member here, at that time a young American from the University of California, Davis, one of the new breed of American scientists who styled themselves as "paleobiologists," not one of the paleontologists of old, in an effort to bring new intellectual vibrancy into the oldest field of Earth science, paleontology, by trying to master two fields, not just one. I had completed two quite different research projects for my still rather newly minted Ph.D., the central goal of which was an attempt to understand how the long-extinct ammonite cephalopods could, after a wildly successful existence on Earth of more than 360 million years, would have gone extinct, while their nearest, lookalike relatives, the still living chambered nautilus, had escaped that fate at the end of the Cretaceous period. I had approached this topic from two different directions, one very nontraditional. The old-school approach was the study of the fossils themselves: anything defective here, any morphology antiquated there, as I examined fossil after fossil over a 20-million-year period prior to their final extinction? Actually, pretty boring work. But the other was a very different approach. Long a deep-water salvage diver of professional skill and experience, I had through chance and fortitude talked my way into a research grant that took me to the one place on Earth where a living nautilus could be actually seen in the wild, the island of New Caledonia, some 700 miles east of the Great Barrier Reef region of Australia. Since that four-month expedition in 1975, I had managed to spend at least a month each year in the water with the wild nautilus and by this time in 1982 had expanded my study area to include Fiji, and I was anticipating with enormous excitement my 1983 field season, already planned for Palau, Micronesia, home to the largest nautiluses (and most beautiful reef walls) in the world. Even the cuttlefish there were giant.

In those years, work with the nautilus was directed by questions that more traditional biologists had never asked of this oldest of cephalopod mollusk, ones that hopefully could shed light on the life span, growth rate, food, and predators of the nautilus that might through inference inform about the ammonites as well, and year by year I arrived in the sunny tropics with better equipment, more grant money, and new ideas and colleagues. But this side of my scientific schizophrenia was increasingly shoving aside geological pursuits, and my presence in Europe in the summer of '82 was not to study fossil ammonites but to look at another living cephalopod that might lend insight into the ammonites, a squidlike animal known as the cuttlefish. This work had drawn me to France, and it was a sheer accident that a chance letter to Wiedmann had led to this invitation to visit one of the few sites on Earth where fossil ammonites could be found in stratigraphic sections with both youngest Cretaceous and oldest Tertiary found in a continuous and well-exposed outcrop.

Wiedmann was definitely old school, a classically trained paleontologist. Sadly enough for the field, by the middle of the twentieth century when Wiedmann had trained in the carnage and chaos of immediate post–World War II Germany, the discipline of paleontology, once a vibrant and necessary area of science important in the study of evolution, had become a sleepy enclave whose every practitioner could spend an entire career writing detailed monographs about the slight differences to be found among the fossil brachiopods of Iowa or among the fossil rodents of Wyoming, studies interesting in their own right but adding very little to larger scientific problems of the time. There were no longer great intellectual questions that demanded the presence of paleontologists at what one eminent British scientist referred to as the scientific "high table." And then, as if out of the blue, a 1980 paper published in *Science* magazine brought the chance of redemption to the field of paleontology, for it was in that year that a group from the University of California, Berkeley, led by a father-son team of Luis and Walter Alvarez, published a bombshell paper forcefully advocating that the K-T extinction was not the consequence of long-term climate change on a multimillion-year time scale but rather was the consequence of a titanic impact of an asteroid with Earth. The Alvarez group proposed two testable hypotheses: that Earth had indeed been struck 65 million years ago by an asteroid estimated to have been 10 kilometers in diameter and that the mass extinction was caused by the catastrophic environmental changes to air and water in the hours, days, and months following the calamitous, really bad day on planet Earth.

What would have killed everything? screamed critics in the weeks following this momentous and eventually paradigm-changing paper. The number of organisms actually killed by the falling rock would have been limited to some few hundreds of square miles. But the surface of Earth is a lot bigger. First the Alvarez group and then others put forth ideas about the actual death mechanisms.

The ultimate killer, according to Alvarez et al., was a several-month period of darkness, or blackout, as they called it, following the impact. The blackout was due to the great quantities of meteoric and Earth material thrown into the atmosphere after the blast, and it lasted long enough to kill off much of the plant life then living on Earth, including the plankton. With the death of the plants, disaster and starvation rippled upward through the food chains.

Several groups have calculated models of lethality caused by such atmospheric change. Apparently a great deal of sulfur was tossed into the atmosphere. A small portion of this was reconverted into H_2SO_4, or sulfuric acid, which fell back to Earth as acid rain; this may have been a killing mechanism but was probably more important as an agent of cooling than direct killing through acidification. However, more deleterious to the biosphere may have been the reduction (by as much as 20 percent for 8 to 13 years) of solar energy transmission to Earth's surface through absorption by atmospheric dust particles (*aerosols*). This would have been sufficient to produce a decade of freezing or near freezing temperatures on a world that, at the time of impact, had been largely tropical. The prolonged winter is thus the most important killing mechanism—and it was brought about by vastly increasing aerosol content in the atmosphere over a short period of time.

Perhaps the most ominous prediction in this model is the formerly unappreciated effect that the giant volume of atmospheric dust generated by the impact has on the hydrological cycle. Globally averaged precipitation decreased by more than 90 percent for several months and was still only about half normal by the end of the year. In other words, it got cold, dark, and dry. This is an excellent recipe for mass extinction, especially for plants—and the creatures feeding on plants.

How to test this hypothesis? More sections of K-T age had to be studied, and those studies had to go in two very different directions. First, geologists specializing in geochemistry had to ascertain if mineral and chemical samples from thin "boundary" layers showed the same kinds of evidence that had first led the Alvarez group to this sensational report. But secondly, the fossil record prior to those beds containing evidence of the catastrophe had to be studied, and studied in far greater detail than had been done before. It was pretty intuitively simple what the fossil record should look like as a result of an impact: There should be lots of fossils at constant diversity right up to the impact layer—and then a vast disappearance of both individuals and species should be very obviously appearing. But the Alvarez team contained no paleontologists. And thus paleontology was given an unexpected pass to the "high table" in one of the most important discoveries of any science ever. One of the greatest questions was as follows: The sections studied by the Alvarez group were found in Italy, near the town of Gubbio. The beautiful limestones making up these rocks had been deposited on a quiet, deep seabed. But the very depth of the water meant that deposition took place in an underwater environment that had few larger animals living above, on, or in it. This deep, black sea bottom had at most a sea urchin or two. What it did have in abundance were untold numbers of microfossils, mainly from two groups. Specialists showed that the fossil records of foraminifera and coccolithophorids showed the predicted pattern of sudden extinction. But because no larger fossils—such as the all-important ammonites—

existed in these rocks, the major question as to whether the impact, if it happened at all, had killed off the more celebrated of the larger marine animals, from ammonites to clams to fish, to the largest marine reptiles such as mosasaurs—let alone the most iconic Cretaceous inhabitants—the terrestrial dinosaurs—could be answered only through the study of other sections. A huge opportunity was presented to the paleontologists. As it turned out, in the majority of cases the paleontology community was not up to the challenge. The paleontologists who studied vertebrate fossils were the most vehement in their opposition, and ironically, the leader of the anti-Alvarez forces was vertebrate paleontologist William Clemens, a specialist on the last dinosaurs in Montana who, like the Alvarezes, worked at Berkeley.

The search was on for stratigraphic sections, places where piles of sedimentary rock of latest Cretaceous and earliest Tertiary age, could be studied. The most useful of these would be sections with the largest variety of fossils available. As it turned out, some of the best of these in the world were the Basque seacoast cliffs. Wiedmann was the geologist with the most experience in these rocks, and through this twist I held keys to important questions. And since paleontologists are very territorial about their established field sites—more so than practitioners of other fields are of their own—Wiedmann found himself in a rather enviable position. Thus, my excitement was enough to help me talk my way into a tour of the most important of the Basque sites, the seacoasts at Zumaya.

We parked high above the ancient Zumaya town square, geared up, and began the quarter-mile hike along a narrow sheep path leading to a steep stairway giving access to Zumaya's rocky beach. These stairs had been cemented against—and in some places carved out of—an enormous bedding plane of—sedimentary rock, hundreds of feet to a side, an originally flat sheet of strata deposited on a deep bottom 66 million years ago but thrown up some lesser millions of years ago as a consequence of the tectonic formation of the Pyrénées mountain chain and now rakishly tilted skyward. Back in the Cretaceous, when this huge stratal surface was but a tiny part of a sea bottom covering the oceans of the entire world, its limy bottom had enough internal consistency that every movement of the varied invertebrates left a trail in the sediment, eventually cemented to form what is known as trace fossils. Worms, crustaceans, echinoids, starfish—all moved across the bottom on a day of their daily lives, and while perhaps none ever made the immortality of body fossil preservation, their behavior was preserved, a testament to the geologists of just how alive that ancient Mesozoic world was before its sudden end, and a stark reminder as well of how few are the kinds of animals that leave fossils, shells, or bones behind at all. The stairs passed down across this track-marked stratum, a painting of a long-gone world.

At the bottom of the stairs, we headed north, scrambling over the wet and slippery maroon strata, more than once slipping into the waiting sea or tide pool, barking shins or scraping skin on the razor-sharp barnacles in the process. But the pounding surf on the rocky points, the scudding clouds, and the vast cliffs that echoed back the crashing of waves on rock vastly overawed these temporal nuisances as we scrambled up and over stratal ridge after ridge, each several-inch to several-foot limestone layer representing 24,000 years, the limestone alternating with darker shale and all controlled by orbital cycles first discovered by a Russian named Milutin Milankovich. The last rocky point, made up of several dozen of these couplets, was the most difficult of all to get over, for like the huge stratal sheet with the stairway, it was tilted about 60 degrees from horizontal, too steep to climb, too steep to safely slide down, and here there was no providential stairway built by obliging Basques. Lowering ourselves hand over hand, the last 10 feet an ignominious slide into a cold tide pool at the bottom of the stratum, a now thoroughly wet duo at last stopped to admire the grandeur of what earlier geologists had aptly named Boundary Bay. Huge walls on three sides enclosed the large bay with a flat, rocky bench about the size of a basketball court exposed at lowest tide, in the rear of the large box canyon, a strand completely water covered at high tide. It was like being in a huge cathedral where the roof and one wall had been taken off, the sheer wall-like cliffs rising a hundred feet or more above the small beach, each wall brightly colored as if painted by some giant. The rocks to the south were a deep maroon in color, those to the north a brilliant white and pink striping. And in the center of the back wall of the bay there was a meeting of the two different units, a sudden transition from maroon beds below to pink and white beds above, starting near the sea and then rising upward from the base of this canyon as the tilt of the beds carried this K-T boundary layer, one the year before discovered to be packed with all the hallmarks of the K-T impact itself, the diagnostic iridium, shocked quartz, and glassy spherules, all save the iridium originally Mexican inhabitants that were now on permanent vacation at this beach (and at all other K-T boundary sites as well over the entire globe).

We walked to this boundary, made up of about a foot of dark clay sitting in between the much more gaudily colored rock layer of before and after. The dark clay seemed an ominous marker, but in reality it was an aftermath, not the calling card of the extinction itself. The rocks above, the rocks below, both were light in color, and that lightness came from the skeletons of untold numbers of calcareous skeletons that had been secreted by microscopic, floating algae in the long-ago latest Cretaceous and earliest Tertiary oceans. So abundant were these tiny plants, known as coccolithophorids, that their dead settling skeletons painted the ocean bottoms a bright white, accumulating over the eons into thick white rocks—the familiar chalk. The chalk seas flourished before the extinction, and after, but not right after. For tens of thousands of years after whatever caused the extinction, the chalk was nearly gone, death removing it from the seas. And in its absence the only sediment grains reaching the seafloor were small grains of rock eroding from the nearby land areas. It is this dark rock, bereft of chalk skeletons, that made up the foot of dark clay, called the boundary clay.

But there was a final layer to reckon with here, the cause of the entire ruckus, much thinner even than the clay layer. We scrunched down, knees complaining as we knelt on the wet rock beneath to bring our heads within inches of the uppermost Cretaceous chalk layer. I pulled out a hand loupe, its ten-power lens briefly sending a moving spotlight of bright light across the outcrop, like a balcony searchlight moving from stage right to pick out the star of the show. In the lens a thin, red layer of rusty-looking grains grew bold. This layer, an eighth of an inch thick, was filled with small spheres of glassy material, as well as small fragments of rusted metal. But hidden in this layer at even smaller size were metals even more rare than iron on a Spanish beach: tiny grains of platinum and iridium, the stuff of stars and the asteroids that circle them. Such a thin layer to cry out that a world had ended in a crater ejecta bombardment, producing fire and subsequent acid rain.

The extraordinary thing, not yet known then in 1982, was not that this layer sat there sandwiched between vast piles of chalk layers, a doomsday special of the epoch. No, the extraordinary thing was how similar this layer looked to others even that summer being examined by other geologists, at places in Europe named Caravaca, Agost, El Kef, Sopelana, Bidart, Stevns Klint, all places where other thin impact layers marked the end of so many kinds of marine life. And it was not just Europe. The impact layer was eventually to be found in marine strata exposed in Russia, the Crimea, Georgia, along a long area of the Black Sea, all the way to Japan and New Zealand; it was found along the east coast of North America, into the Caribbean, to South America, all the way to Antarctica. It even looked like layers found from land-derived strata, in places such as Hell Creek, Denver, and Judith River; up the Milk River regions of Alberta, and far into the Arctic. That was the most salient fact learned from this event: Its calling card was global and easily recognizable. There was a vast replication of this sequence, and as geologists fanned out over the years to study these places, there came a vast, comforting (for all but the ever-diminishing ranks of naysayers) confirmation through replication of fossil records terminating at chemically similar layers increasingly believed to have been caused by the rainout of the vast crater carved into Earth, 65 million years ago.

We placed hands on the boundary clay, as if expecting some message from the dead, but there was nary a peep, so we began to work. Starting at the point of catastrophe, with a meter stick, we began to slowly measure the thickness of each bed, each number scratched into a yellow field book; down section and thus back in time we went, bed by bed, hour by hour. I was amazed as the German pulled out a spray-paint can and painted gaudy Teutonic numbers on the rocks of large and ugly size, marking each successive 10-meter interval of strata beneath the K-T layer. Soon the tide began to rise, but by then we were already out of Boundary Bay, repetition breeding more speed, but the tide was not to be denied, and long before finishing even the 100 meters of stratal thickness between the K-T boundary and the stairwell, we were forced out of the bay by the rising water. But the framework for our morrow's work was in place. Any fossil collected in the succeeding two days would be collected from a layer of a known distance below the death layer.

It had been on the long drive of the day before that I had asked the ever correct, pleasant, but distant German professor if the collections from this place made over the many years of study by him and his yearly group of spring-semester students could be used to test the second of the two hypotheses proposed in the 1980 Alvarez paper: that the fossil record should show a catastrophic appearance, with many species disappearing suddenly from the succession of beds in the layers just beneath the thin impact layer. Wiedmann pondered for a moment. "I doubt it," he said. All of the hundreds of ammonites collected over the decades were simply labeled as coming from Zumaya. But it was his strong recollection that the ammonites disappeared gradually, not suddenly, because that is how mass extinctions worked. All were gradual.

I was silently astounded. Men such as Wiedmann had been my heroes in grad school, and my own major professor had been Wiedmann's fellow grad student in post–World War II Germany. Wiedmann himself had become the greatest expert on the extinction of ammonites through each of the mass extinctions—and more. These were the lineal descendents of the Teutonic, mid-nineteenth-century fathers of biostratigraphy, the great Friedrich von Quenstedt, who had demanded that his followers never tire of the exacting work of collecting fossils from known stratigraphic positions in the vast tables of strata, and his even more brilliant student, Albert Oppel, who pioneered the use of fossils to produce the finest division of time possible, the zone. Wiedmann, their heir, one of the godlike German professors of paleontology, had apparently tired of the exacting work.

From the first announcement of its discovery in 1980, and then continuing well into the 1980s, the Alvarez team exhorted paleontologists to test its groundbreaking hypothesis using fossils. To do that, many different K-T boundary sections would have to be studied, and Zumaya looked like a perfect testing ground for this most compelling of scientific hypotheses. But it looked as if all new collecting had to be conducted. The entire section had to be measured, and whenever a fossil was found, it would have to have its level in meters below the impact layer exactly noted. With enough collecting in this way, one of the major predictions of the Alvarez impact hypothesis could then be tested: Did the ammonites disappear suddenly or gradually? If many species and individuals were found just below the boundary, it would be evidence of sudden extinction. But a long, slow diminution would be a major blow against Alvarez et al.

At the end of the two days, the approximately 50 fossils were pried from the stratal walls, each from a bed of known distance below the boundary. Ultimately they were never to play any part in the controversy, for a controversy was what this had become in the early 1980s. But one thing came through this first collecting attempt by the young America and older German. Try as we could, neither of us had been able to find an ammonite within 15 meters of the impact layer in Boundary Bay, and both of us had come into our field with the ability to find a fossil when no one else usually could. But not this time.

Wiedmann seemed very pleased. Whenever I brought up the Alvarez hypothesis, Wiedmann was wont to mutter a deprecation in German. Sudden extinction? Ridiculous. This was becoming the knee-jerk reaction by all but the youngest (or really good older) paleontologists worldwide. To these men (there were indeed very few women in the field in the immediate post–World War II generation), there was no way a mass extinction could have been catastrophically fast. Catastrophism was a failed nineteenth-century theory, and none of the powerful, mid-career European paleontologists of the early 1980s—and very few of the Americans, either—were going to allow the field to fall back into believing that failed idea. Only geniuses David Raup and Stephen Jay Gould noisily demurred.

Wiedmann packed the fossils into his sporty red Audi (he was newly divorced) and, dropping the American off at a train station, sped off toward Germany and his more important projects, since he was convinced that there was no controversy—while there might have been an impact, there was certainly no rapid extinction among the ammonites.

I made my way across the beautiful Spanish countryside, the leisurely train trip giving me plenty of time to reflect on the experience. Nearing the cerulean expanse of the Mediterranean, I turned back to thinking of cuttlefish, never thinking that I would again see the cliffs of Zumaya. But life has a funny way of changing things, and perhaps it is just as well that we cannot see the future.

THE MASS EXTINCTIONS WERE SUCH LARGE-SCALE AFFAIRS THAT THEY left obvious and indelible records in the rocks, and once an organized way of noting the ranges of fossil in rocks was put into practice, in the early nineteenth century, it became obvious that there had been great catastrophes in the past. But before mass extinction could be recognized, the concept of any sort of extinction had to be proposed and accepted in an intellectual world that for centuries had considered that the creator and his creations were immutable. Once there, they would never go away. It took a great French naturalist to change that.

One of the loveliest parts of Paris is the Jardin de Luxembourg and the adjoining Jardin de Plants. Great limestone buildings line the far end of the park, with busts of the great French geniuses of natural history of the eighteenth and nineteenth centuries gazing emptily down on the flowers and science pilgrims alike. One of them was crucial in founding the science of stratigraphic geology and extinction.

In one of the halls near the edge of the park there is an incredible boneyard amassed by this father of mass extinction research, Baron Georges Cuvier, who was the first to draw attention to the concept of extinction by demonstrating that bones of large elephant-like animals found in Ice Age sedimentary deposits could not be assigned to any living elephant. He deduced that these bones came from an entirely extinct species.

Cuvier's bold assertion was soon corroborated, and in spades. The birth of the geological time scale in the subsequent decades of the early nineteenth century quickly demonstrated not only that species had undergone extinction but also that many had done so in short intervals of time. In order to devise some way of determining the age of rocks, European and American geologists had begun to systematically collect fossils as a means to subdivide Earth's sedimentary strata into large-scale units of time. In so doing, they made the discovery that intervals of rock were characterized by sweeping changes in fossil content. Setting out to discover a means of calibrating the age of these rocks, they also discovered a means of calibrating the diversity of life on Earth. They also found intervals of biotic catastrophe, which were named *mass extinctions*. In a doctrine that came to be known as catastrophism, these were thought to be caused by a succession of worldwide floods or other disasters that killed off most or all species, followed by a reintroduction (or re-creation) of new species.

As the nineteenth century passed into the twentieth, Earth scientists increasingly rejected these catastrophist precepts. But what might have caused these calamities? While mass extinctions were accepted as having taken place, they were viewed as gradual, long-term events, a uniformitarianism view that was held well into the twentieth century. The ultimate cause remained enigmatic, but long, slow climate change—resulting in long, slow extinction—was the favored cant.

The two largest mass extinctions, recognized even as early as the mid-nineteenth century, were so profound that they were used in the 1840s by John Phillips, an English naturalist, to subdivide the stratigraphic record—and the history of life it contains—into three large blocks of time: the *Paleozoic era*, or time of "old life," extending from the first appearance of skeletonized life 530 million years ago until it was ended by the mass extinction of 250 million years ago; the *Mesozoic era*, or time of "middle life," beginning immediately after the Paleozoic extinction and ending 65 million years ago; and the *Cenozoic era*, or time of "new life," extending from the last great mass extinction to the present day. Phillips also made the first serious attempt at estimating the diversity of species present on Earth during the past. He showed that over time, diversity has been increasing, in spite of the mass extinctions, which were only short-term setbacks. Somehow, after each extinction there seemed to be room for larger numbers of species than were formerly present. Far more creatures were present in the Mesozoic than the Paleozoic, and then far more again in the Cenozoic. But the mass extinctions did more than just change the

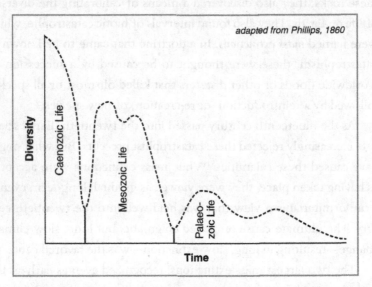

FIGURE 1.1

Diagram from John Phillips (1860:66), illustrating his estimates of diversity of species through time (the present is on the left). Phillips's ordinate corresponded to the number of marine species per 1,000 feet (305 meters) of strata. Notice how the two mass extinctions (Phillips called them "zones of least life") became the means of differentiating the Paleozoic, Mesozoic, and Cenozoic eras.

number of species on Earth. They also changed the *makeup* of Earth (Figure 1.1).

By the 1960s, mass extinction research became dominated by two figures, one German, one American. The German, Otto Schindewolf of Tübingen, is my scientific "grandfather" in a way. He bravely survived World War II without caving in to or joining the Nazi party, and after the war he returned to prominence the great tradition at Tübingen, a place that had been home to two of the great giants of geology, Friedrich von Quenstedt and his disciple Albert Oppel. Paleontologists began to revere Schindewolf for his careful work, and it was he who made the then-heretical suggestion that mass extinctions could have been caused by nonearthly causes, thereby predating the Alvarez confirmation of this idea by two decades at least. Schindewolf worked on many things, but the mass extinctions intrigued him most of all, and he looked at their evidence at many sites. On the American side, on the other hand, another giant also was preoccupied by the mass extinctions but came at them in somewhat different fashion than Schindewolf. Norman Newell of Columbia University began some of the first serious compilations of various extinction rates and for the first time ranked the various extinctions by their deadliness in killing off taxa. (Newell was also in the student-training business, two of his best-known protégés being Niles Eldredge and Stephen Jay Gould.)

Newell classified many mass-extinction events occurring since the "Cambrian Explosion" of 540 million years ago. Yet other mass-extinction events of earlier times are largely unknown to us because they occurred when organisms rarely made skeletal hard parts, and thus rarely became fossils. Perhaps the long period of Earth history prior to the advent of skeletons was also punctuated by enormous global catastrophes decimating the biota of our planet, mass extinctions without record, or at least without a record that has yet been deciphered.

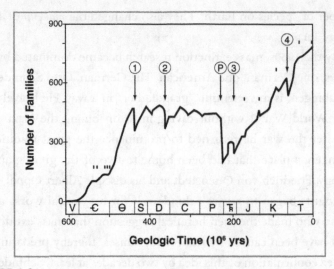

FIGURE 1.2

Diversity through time, as indicated by the number of families found in the fossil record. The so-called Big Five—mass extinctions of the Ordovician, Devonian, end-Permian, end-Triassic, and end-Cretaceous periods—are indicated. The five major mass extinctions: 1. Ordovician; 2. Devonian; P. Permian–Triassic—the "Great Dying"; 3. Triassic–Jurassic; 4. Cretaceous–Tertiary (the K-T). The Paleocene Thermal Event occurred right after the K-T event on this graph.

While Newell began the work on estimating the death rate and continued to labor through the 1960s, this monumental work was taken over by paleontologists David Raup and Jack Sepkoski in the late 1970s, work continuing right through the 1980s and 1990s. Through such statistics the Big Five (Ordovician, Devonian, Permian, Triassic, and Cretaceous) extinctions were recognized. If the number of families going extinct is used for comparison, the P-T (Permian–Triassic) mass extinction leads, with a rate of 54 percent, followed by 25 percent for the Ordovician, 23 percent for the Triassic, 19 percent for the Devonian, and 17 percent for the K-T event. The Cambrian extinctions do not appear as "major," but they were certainly important in reordering life on Earth at the time (Figure 1.2).

By the 1970s it was clear to most—but not all—paleontologists that there were numerous extinctions in the past. Many of those belonging to the group of paleontologists who specialize in the study of humankind's group, the vertebrates, began to distance themselves from the rest of the paleontological community because of profound disagreements over many issues, and the existence of mass extinctions was one of these: Many "vertebrate paleontologists" flat out did not believe that there had been mass extinctions and suggested that the places in the geological record where large numbers of species disappeared in short strata distances were caused by vagaries of the fossils or rock record, not from some real catastrophe. This group began to meet separately from the other paleontologists (the micropaleontologists, paleobotanists, and invertebrate paleontologists), and when the Alvarez findings were published, it was from this group that the loudest dissent and opposition came. To the others in the fossil fields, however, the evidence that there had been these die-offs in the past seemed overwhelming. But what were the causes of these events? Could all have been the results of one kind of cause, repeating itself through time, the way the Black Death returned to medieval Europe every few decades, or were there as many causes as extinctions?

Before cause could be ascertained, it first had to be learned how similar in terms of rate and breadth of dying the events were, and quite quickly two very different kinds of mass extinctions were posited, differentiated by the rate of dying. A "gradual mass extinction" would have been characterized by a slow reduction of species over some period of time—that is, species would have been smeared out over some extensive stratigraphic interval. Long-term climate change has been cited as a cause of this type of mass extinction. The second type, "catastrophic extinction," or "rapid mass extinction," would have been characterized by disappearance (extinction) of species over a short period of time, or stratal interval.

Prior to 1980, all of the mass extinctions were thought to have been of the former type. And there was a second and largely overlooked aspect of the "science" of mass extinction research prior to 1980. None of the hypotheses for the past mass extinctions—such as slow climate change, disease, lowering oxygen, changing sea level, increased predation—were testable. But all of these possibilities seemed reasonable, and all could be seen to be a way to gradually kill off species. Not so for the rapid extinctions. While a rapid mass extinction could be theorized, there seemed no possible terrestrial mechanism to provoke one. But when the theorists began thinking outside the box, with the box being Earth, a number of possibilities came to mind.

Even before 1980 it seemed pretty clear that a large enough asteroid impact would cause a very rapid extinction, when seers such as David Raup of the University of Rochester and the great Digby McLaren of the Geological Survey of Canada proposed that ancient impacts might have caused some of Earth's past mass extinctions. Raup was even modeling how impact could cause extinction as late as 1977, only three years before the publication of the paradigm-changing Alvarez paper. Using simple computer programs, he simulated the effects of asteroids of various sizes hitting Earth. Bigger or smaller, hitting this continent or that ocean, Raup watched as his program scythed through Earth's biota. He was fixated on asteroid bombardment as a cause of past mass extinctions, one of the perennially hottest of paleontological topics, because the extinctions played so large in the geological and evolutionary past, and ironically, and unknown to him as he worked on his computer programs, a group of scientists studying rocks brought back to Berkeley from the mountains of Italy were about to make one of the greatest discoveries of any science. Strangely, Raup never published a paper on his computer results, perhaps because of the (then) lack of evidence that any past impact had done anything biologically to the denizens of Earth's past. Was that about to change!

It is safe to say that the 1980 Alvarez et al. paper turned not just paleontology on its head but also the entirety of Earth sciences, as well as a large hunk of evolutionary thought. So much has been written about it that it will get short shrift here, not so much because of any lack of importance but simply to avoid rhetorical overkill. But in the tradition of scientific paradigm change promoted by philosopher of science Thomas Kuhn, the Alvarez paper certainly was the first shot of a major scientific revolution. (Kuhn suggested that areas of science are organized under large-scale paradigms. There is much science done under such a paradigm, and most of it simply further reinforces the big scientific tent that a paradigm might be analogized to. But every once in a while new information knocks down the tent poles, and there is a period of revolution until a new and different tent is erected.)

In the case of mass extinctions, the major paradigm was that all were slow, lasting millions of years for their major transition from, say, Paleozoic life to Mesozoic life. And the major theorized cause was slow climate change. There were two poles to the mass-extinction paradigm, then: They were slow, and they were caused by Earth-bound conditions. The Alvarezes proposed that neither of these were right.

The passage of what might be called the Alvarez impact hypothesis from controversial paper to accepted scientific fact is one of the great studies not only in how scientific paradigm change but in human nature and behavior as well. The bigger the paradigm change, the bigger the stakes for supporters of each side. In this case, there was not one but two very different (if causally connected) hypotheses involved, and a third that was implicit. The first, it turned out, was within two years supported by so much data that it was almost universally accepted as fact. The second, however, that the effects of the impact caused the K-T extinction, took longer to confirm. This was essentially due to the very different nature of the data that had to be collected. The third, implied hypothesis was that if the K-T event had been caused by impact, then some (maybe all!) of the other mass extinctions had a similar cause. Let us look at each of these hypotheses.

First, there was the hypothesis that an impact occurred. Two of the most important lines of evidence used to convince most workers that the K-T impact layers were indeed caused by large-body impact was the discovery of both elevated iridium values within the boundary clays and abundant "shocked quartz" intermingled with the iridium. These were quartz grains that showed multiple thin lines called shock lamellae. Most recently on Earth they have been produced on small sand grains by the explosion of nuclear warheads during underground testing. They are also found in meteor impact craters; no conditions on Earth naturally create such quartz grains with multiple shock lamellae.

Another characteristic of the impact layers was large numbers of beadlike glassy spheres, smaller than a millimeter in size at most sites. These spherules resembled tektites and were interpreted by the impact group to have been formed by earthly material blown into space during the impact, only to return to Earth. But in the return, these bits of tiny rock melted to produce glass spherules, which eventually hit the ocean and settled onto the bottom amid other material deposited after the impact, such as the shocked quartz, and tiny bits of iridium.

In addition to iridium, shocked quartz grains, and spherules, the thin K-T boundary impact layer sites also ultimately yielded evidence of fiery conflagration that must have occurred soon after the impact. Fine particles of soot were found in the same K-T boundary clays from many parts of the globe. This type of soot comes only from burning vegetation, and its quantity suggested that much of Earth's surface was consumed by forest and brush fires.

By 1982, high iridium concentrations had been detected at more than 50 K-T boundary sites worldwide. Thus, early on, the geochemical evidence found at ever more K-T impact sites rather quickly changed skeptics into supporters: first the enhanced iridium, then the spherules, capped by the shocked quartz grains—three indications that a rock not of Earth threw lots of Earth rocks back into space for a short time, only to have them fall back from the sky to coat the entire surface of the planet, every square inch, with this patina of Earth and stardust. And as if this was not enough, soon those who studied geochemistry brought forth yet another type of evidence that an extinction followed very soon after the impact—very soon indeed.

Carbon is, of course, one of the most important of all elements on our Earth. It is found in a range of minerals and rocks, but it is an important constituent of life itself. It turns out that when extinctions occur, the movement of carbon atoms from the living to Earth, and back again, is changed. Early in the twentieth century new generations of machines called mass spectrographs enabled geologists to better track the movement of carbon in and out of the ocean, Earth, the atmosphere, and life itself. This movement, called the carbon cycle, is now well known through years of study. One of the more interesting discoveries about the movement of carbon through time is what happened to it during the great mass extinctions. By taking small measurements of various sediments or fossils for its carbon content, it was found that important clues to the rate and cause of mass extinctions could be gleaned. Here is how that works.

Carbon atoms come in three sizes, or isotopes, with slightly varying numbers of protons and neutrons. Carbon-14 (^{14}C) decays at a rapid rate that is often used to date particular fossil skeletons or samples of ancient sediments. But for interpreting mass extinctions, a more useful type of information is the ratio of carbon-12 (^{12}C) to carbon-13 (^{13}C) isotopes, which provides a broad snapshot of the types of life predominant at the time. That is because changes in the ^{12}C:^{13}C ratio are largely driven by photosynthesis: Plants use energy from the sun to split carbon dioxide (CO_2) into organic carbon, which they exploit to build cells and provide energy, and happily for us animals, free oxygen is their waste product. But plants are finicky, and they preferentially choose CO_2 containing ^{12}C over the slightly larger (by one neutron) ^{13}C isotope. As a result, a higher proportion of CO_2 remaining in the atmosphere contains ^{13}C when plant life is abundant on Earth—whether in the form of photosynthesizing microbes, floating algae, or tall trees—and atmospheric ^{12}C is measurably lower.

But plants are not the only organisms that employ CO_2; the formation of a clamshell, for instance, involves the precipitation of calcium carbonate, requiring carbon atoms. Clams are far less picky and use both isotopes, but if a mass extinction had swept away most plant life, thus reducing photosynthesis, all clams in the new, deader world would have encountered a greater supply of ^{12}C. This information is incorporated into their skeletons, and by collecting a series of such samples from before, during, and after a mass extinction, investigators can obtain a reliable indicator of the amount of plant life both on land and in the sea.

For the K-T event, the carbon isotope curve shows a simple pattern. Virtually simultaneously with the emplacement of the impact layer containing the impact debris (the iridium, shocked quartz, and glassy spherules), the carbon isotope pattern shifts—more ^{12}C is present relative to ^{13}C—for a short time, and then returns to its old, pre-impact values. This makes sense if a large amount of Earth's plant life, both on land and in the sea, was suddenly killed off, was dead for a while, and then came back to life. And it is entirely consistent with the fossil record of those two groups: Both larger land plants and the sea's microscopic plankton underwent staggering losses in the K-T event.

This indicates that for a short period of time, there must have been a worldwide and devastating extinction of plants. Not only were most species killed off, according to these new data, but also perhaps the majority of individual plants themselves. The reason is not hard to find. Soon after the impact, most of the forests burned to the ground, and those plants not killed by that conflagration were then subjected to massive changes in temperature and water availability. Under a blanket of cloudy debris from the smoldering crater, Earth cooled for decades, and the tropical vegetation of the steamy, hothouse Cretaceous period largely froze to death. It was a single, neat record: bang, change, return to normalcy—except that most of the plant and animal species characteristic of the pre-impact period (dinosaurs, ammonites) were gone.

All of this evidence provided comfort to the Alvarezes (and the legions of scientific supporters and media supporters they had by then). Was there ever a more news-friendly science story? Dinosaurs, death, asteroids, everything but alien sex. But never count out foes who just cannot afford to lose—massive reputations, massive egos were at stake. In the mid-1980s came a determined counterattack by the nonimpacters. While no one now doubted that the K-T impact layers existed, a group of geologists, led by Charles Officer and Charles Drake, proposed that large-scale volcanism could have produced the impact layers. They pointed out new studies showing that small but significant amounts of iridium could be found emanating from active volcanoes on Hawaii and explained both the shocked quartz and glassy spherules as being related to volcanism, not impact. Finally, they had another very interesting bit of information to use as argument.

At about the same time that the K-T extinction took place, a large area of what was to become India slowly became covered with lava, eventually, through its very size and area covered, becoming a "flood basalt." Many such flood basalts are visible on Earth today, in addition to the large stacks of lava, shown to be slowly accumulated over about two million years. For instance, in parts of Washington State, Oregon, and Idaho, an enormous area of land is covered by black basalt many hundreds of meters thick. All of this lava must have oozed out over the land to eventually cover hundreds of square miles. It came not from a succession of volcanic cones but from great cracks in the land itself. Such flood basalts produce more than just lava on land (or under the sea, for they can occur here as well): As the runny magma rushes out into the air from its deep Earth origin, it carries enormous volumes of volcanic gas into the atmosphere. These gases include toxic components, such as hydrogen sulfide, as well as methane and, perhaps most important—carbon dioxide. If flood basalts are combined on a global scale with more explosive volcanism, the kind that throws great quantities of ash and volcanic dust into the atmosphere in addition to the volcanic gases, one might expect major effects on animals and plants. This reasoning became the major competing hypothesis to the Alvarez impact theory.

In a series of scientific meetings over the decade, the impact and volcanism sides met face-to-face, presenting their respective data accumulated unusually simply to support an already decided view. But it became clear that as the decade progressed, the impact group, supported at first by massive confirmation of the makeup of impact layers across the globe and later by an increasing number of paleontological studies showing data consistent with a sudden extinction, "were opposed by" the ever-decreasing doubters who were increasingly composed of cranks, the slow and conservative, and those seeking attention by screaming in loud if knowingly false protest.

It became increasingly clear that there could not have been enough volcanoes on Earth to have produced the amount of iridium found in the K-T impact sites. But in one area, the volcanic side had found a relationship between volcanism and mass extinction that could not be shouted down by the impact side and left the more introspective among the impact camp feeling rather uncomfortable, although few would admit to as much. In an increasing number of studies, geologists using new dating techniques to look at the ages of Earth's largest flood basalt provinces were surprised at how large some of these provinces were.

The Columbia River Basalts of the Pacific Northwest, for example, are staggeringly large to those who must drive across them in any trek from Seattle to Spokane or Idaho. And yet it was discovered that the Columbia River Basalts are small for flood basalts. The K-T debate stimulated new research into the flood basalts, and a surprising result was discovered: The largest (volume) of them seemed to closely correspond in age to the times of each of the great mass extinctions of the last 500 million years.

The largest flood basalt of all, named the Siberian Traps (and they are indeed in Siberia), was deposited over the same time interval (around 252 million to 248 million years ago) as the most catastrophic of all the mass extinctions—the great Permian extinction of 251 million years ago. A second giant flood basalt, mainly underwater in the central Atlantic, but also underlying the Brazilian rain forest, was named the Central Atlantic Magmatic Province, and its age—202 million to 199 million years—again closely corresponds with the Triassic mass extinction of 200 million years ago. The list goes on and on. Even small extinctions, such as that at the end of the Paleocene epoch, some 60 million years ago, corresponded to a flood basalt.

This very curious finding led some to propose a hybrid of the two hypotheses—that the impact of an asteroid so shook Earth that it unleashed a flood basalt somewhere on Earth's surface. Even when astute geophysicists showed time and again that such a cause could not produce a flood basalt effect—the idea never would go away, and newer versions of it have appeared as late as 2005. Eventually the impact camp simply shrugged this all away as coincidence.

Part of the reason that the acceptance of the first part of the Alvarez impact hypothesis—that Earth was hit by an asteroid at the end of the Cretaceous period—was so quickly achieved is that all of the necessary sampling was confined to the impact layer itself, and at most no more than a meter or so of strata both above and below the layer. In a single day the geochemists could go in, dig out their samples, and be finished. The paleontologists, on the other hand, had a very different set of problems that inherently required a far-longer interval just for sampling. It takes longer to sample for fossils, and the larger the fossils, the fewer there are. For fossils the size of ammonites, it turned out that multiple seasons, not days, were required to accumulate sufficient numbers of data points to allow any sort of meaningful analysis. Very few good paleontological studies were available. Thus, for the second part of the Alvarez impact hypothesis that the extinction itself was caused by the impact, there was far less acceptance, at least among those best trained to make a meaningful decision. It took much longer to study the fossil record at the K-T boundary than to simply dig up the millimeter-thick impact layer at the boundary itself.

The ammonite fossils from Zumaya eventually did play a large role in supporting the contention that the K-T mass extinction among not only microscopic marine plankton but also among macroscopic animals living in the latest Cretaceous oceans was caused by the impact. It took a while, however. As it turned out, it would require three field seasons at Zumaya to accumulate enough ammonites to deduce anything meaningful, and eventually it was found that no amount of collecting from the highest beds at Zumaya, those just beneath the impact layer and thus the most critical for testing whether ammonites were there for the last dance, was enough simply because of the impossibility of finding any fossils in the highest beds because of their orientation. But that is getting ahead of things.

WITH JOST WIEDMANN'S BLESSING, I RETURNED TO ZUMAYA LATER THAT summer in 1982, and for a much longer collecting trip in 1984. Enough

ammonites were collected to allow us to publish a paper in 1986 showing the diversity of the ammonites approaching the boundary, and indeed even this new collecting still supported Wiedmann's view that
the ammonites were dying off before the boundary and hence were
likely not driven to extinction by impact. At best, according to Wiedmann's interpretation of these data, the one or two species still around
at the time were killed off. But more than 20 different species were
known from the approximately 100 meters of strata from the entry
stairs to the boundary at Zumaya, and again our highest ammonites
came from about 15 meters below. But the nagging problem, at least
to me, was that these last 15 meters were oriented in such a way that it
was impossible to see if any ammonites were enclosed in these strata.
And by this time I had found a small fossil cephalopod right beneath
the boundary at Zumaya. But it was too small to be identifiable as
either an ammonite or one of the group of fossils that we knew survived, the nautiloids.

Elsewhere along the vast coastline, ammonite fossils could be
found by looking at either the upper or the lower side of a sedimentary
bed. The beds themselves were tilted, and the bay itself was formed
by enormous bedding plane surfaces, where literally hundreds—
perhaps a thousand—square feet of perfectly cleaned and oriented
beds allowed the discovery of any ammonite enclosed. In Boundary Bay, however, only the edges, never the tops or bottoms, of beds
could be seen. Perhaps the lack of ammonite fossils there was only
collection failure. Nevertheless, our 1986 publication of the Zumaya
ammonite data was like a life raft to the drowning opposition to the
Alvarez supporters, and I found myself lionized by the opposition to
Alvarez, even though I endlessly tried to explain the possibly mitigat-
ing circumstances of the collecting. That was like whispering to a faraway companion in the middle of a storm.

To finally come to some decision about the extinction of the ammonites in the Basque Country, we needed strata ideally positioned to allow collecting over the past 15 meters. Happily enough, a new trip back, in 1987, revealed two such places, both as rich in fossils as Zumaya but positioned on the coastline in such a way that the crucial, final layers of strata leading up to the boundary could be searched for ammonites.

BIDART, FRANCE, SEPTEMBER 1987

After the three collecting trips to Zumaya, the first foray out of Spain and into France was like a breath of fresh air. The Spanish Basques are a dour lot, and the language barrier (the Basque language is a very tricky one indeed) made any sort of friendly social interaction impossible. Worse yet, they did not open the restaurants for dinner until 10 PM, and after an entire day on the rock, climbing up and down the stratal sheets, hacking fossils out of the cliffs, and hauling rock samples long distances, a young man is hungry by 6 PM. The move to France, first made in September 1987, was an alimentary godsend. And, more important, the rocks themselves were tilted in our favor.

Zumaya is approachable only by the long trail from town, and on our several trips we geologists almost always worked in complete solitude. The two new localities in France, each with magnificent K-T boundaries, were either on or accessed by large sandy beaches that were major tourist destinations. On a sunny afternoon, and most afternoons were sunny during the summers when I was there, very little sand showed from beneath the packed, tanning, and largely nude thousands of Europeans. Oddly enough, I never encountered a single American there.

Map exploration led to a K-T section in a beautiful high park just north of the Spain–France border at a place named Hendaye, a place with an odd juxtaposition: An ancient castle was surrounded by Hitler's

Atlantic Wall pillboxes, seemingly guarding the strata containing the K-T boundary. A second locality was marked by the French geological maps as existing on a beach near the town of Bidart, some 15 miles up the coast from the Hendaye and some 60 miles north of Zumaya.

After scouting the Hendaye site, which, like Zumaya, was at the base of rocky cliffs, I made my way up the coast to where the Bidart beach should be. (Unfortunately, unlike Zumaya, at Hendaye there were no providential stairs hacked down the hundred-foot cliffs to get to the outcrop, and eventually my various teams and I had to hack our own path down out of the rock.) Car parked, I followed a path through sand dunes down toward the sunny, surf-swept strand.

The Cretaceous and Tertiary strata turned out to be at the back of a wide, sandy beach, the kind of sand that demands bare feet and a release of all other plans, thoughts, or anxieties. Thousands of adults and children frolicked in the waves or in the surf or simply baked in the sand, this being well before the skin-cancer scares, and a tan was a necessary accoutrement to any sharp Frenchman or Frenchwoman. The other surprise, after coming from the conservative Basque enclaves (and the conservative American beaches as well), was that the most promising K-T boundary section had been commandeered by a large contingent of obviously gay Frenchmen, if hand-holding, kissing, and furtive climbing back into the sand dunes from time to time by groups of two or more men was any indication.

I must have been a pretty ridiculous sight: a strangely garbed person climbing up and down the moderately high cliffs, smacking and lustily digging into the soft Cretaceous sediment (for that indeed was what it was—unlike the strata at Zumaya, where nearby Pyrénées mountain building had hardened the deep sea strata finally brought up from its deep burial, the Cretaceous and Tertiary chalks at Bidart were soft enough to allow endless beachgoers to carve the strangest graffiti in French, German, and Spanish, using pen knives. Our rock hammers made short shrift of the stuff). Being the only clothed (and untanned) person amid perhaps a hundred lounging, preening, flirting, and completely nude men of quite variable ages was a strange experience. But each whack of the geological hammer sent rock chips flying in all directions, and this actually turned out to be a saving way of keeping some slight distance of personal space between the curious beachgoers and the working geologists.

If less changed by burial and heat over the long roll of geological time, the succession of strata making up the Bidart beach section was far more affected by faulting than the Zumaya rocks had been. More than 200 meters of beds piled one on another could be accurately measured at Zumaya. But at Bidart, major faults had thrown the strata this way and that and chopped what had been a continuous section into small packets. But the most important region of strata, that containing the K-T boundary itself, contained a precious 20 meters of continuous beds, one atop another, right up to the beautifully exposed impact layer. All was ready to really test Alvarez.

It was at sunset that I began seriously collecting around the boundary. A half dozen men lay around on the golden sand in the small regions I was sampling, the nearby sea sparkling with sunlight as the long Atlantic rollers foamed into whiteness, then disappeared into the sandy strand. A perfect afternoon. Perfect rocks. A loud hoot from the clothed man as the first ammonite was found, within a meter of the boundary, a louder one as an even better specimen revealed itself within 20 centimeters of the boundary layer. Specimen after specimen, and not only ammonites in abundance, but in diversity as well. Eventually a dozen different species would be found in the last meter of strata, and my eventual scientific papers corrected the earlier findings that ammonites died out well before the boundary. In fact, they showed absolutely no change in abundance and diversity until the very bad day that they were killed off by the great asteroid.

So it was not only microfossils killed off by the impact. With this volte-face I was dropped from the conclaves of the opposition to impact and embraced by the impacters themselves, including their head honchos, because of the importance of showing that animals as well as microbes were susceptible to meteorite fall. I eventually received the warmest letter of thanks from Walter Alvarez soon after the publication of these data, which was nice in light of the far ruder things his father Luis had said about me a year earlier, when the quite different (and far less Alvarez-friendly) results had come out.

I had enough fossils by the end of this field trip from the highest beds of Hendaye and Bidart to show quite conclusively, at a meeting held on the October 1987 day that the American stock market so precipitously fell that the ammonites were indeed victims of the impact. The meeting had attracted many of the key players in the extinction game, including Jost Wiedmann. By this time summer was long gone, and I was nearly deathly ill with bronchitis from nonstop work in the autumnal gales, having been some two months away from home for this joyful acquisition of fossil data.

When it was time for my talk, Wiedmann moved to the front row and listened intently. He had just presented a paper in which he reasserted that no animal could have gone extinct in any sort of asteroid impact, since (1) there had not been an impact and (2) even if there had been one, all the ammonites were already extinct before the end of the Cretaceous period. As I presented slide after slide showing ammonites not only present but thriving up to the boundary, he turned increasingly pale. At the end of the talk he slowly walked out the door and headed to his car. He sped off, and we were never to speak or even communicate again.

He died several years later, a sad scientist whose life work was shown, at the end of his life, to have been quite wrong, with insult added to injury by having this demonstrated by one of his own apprentices.

—————

MINE WAS BUT ONE OF MANY STUDIES APPEARING IN THE LATTER PART of the 1980s and into the 1990s that confirmed what the much smaller fossils had indicated years before: The extinction was short in duration. Regarding marine mollusks such as my ammonites, terrestrial vegetation, and eventually even the terrestrial vertebrates, including the largest of all, the dinosaurs, it became clear that the many earlier studies suggesting a gradual diminution of diversity millions of years before the K-T were false. It was learned that vagaries of sampling or preservation rendered almost all of the early studies, such as that by Wiedmann and me, just plain wrong.

Though late, the paleontologists finally limped into camp with confirmation of the second part of the Alvarez hypothesis, that the impact did indeed cause the mass extinction. And from this we came to understand what impact extinction acted like. It was like an earthquake hitting a city. One moment everything is normal, the next all is calamity. Like the complex patterns of physical and social interrelationships in any big city suddenly if not totally destroyed, at least very much so, those not killed in the shake may later fall victim to the massive perturbations or destruction of water lines, power, food acquisition, shelter, and social order and the rapid spread of disease. One day the Cretaceous world was living its life; the next it was destroyed, the deaths coming either that day or in the weeks, months, and even years afterward. But for both city and living planet hit by asteroids, one thing is sure: Recovery starts immediately. When the aftershocks finally still, the rebuilding starts.

AS THE 1980S CAME TO AN END, THE DEBATE ABOUT THE K-T EXTINC-tion changed from one characterized by collegial exchanges between ever-hardening sides to one of humorless, angry polarization. There were two sides and no middle, and by that point, no prisoners were being taken. The impacters increasingly lorded it over the volcanists. Because almost the entire geochemical establishment was impacters, while so many of the nonbelievers now allied with the volcanists were paleontologists (and especially vertebrate paleontologists), things became uglier yet when all members of an entire scientific field began to belittle another one.

It did not help that the belittler in chief had a Nobel Prize and was also the senior author of the momentous 1980 paper that started the entire thing. Luis Alvarez was increasingly frustrated that so many paleontologists still refused to accept that the extinction was caused by the impact. (They gave him his impact at least but held the highly dubious stance that it was a coincidence that the largest impact of the last 500 million years exactly coincided with one of the five largest mass extinctions of that interval). He began to refer to paleontologists as "stamp collectors." He was especially cruel to William Clemens of Berkeley, and as the years went by, Bill, an old friend of mine (who became much less of a friend as my ammonite data began to appear in the literature), aged quickly.

To the winner go the spoils (academic honors, which lead to higher salaries); to the losers goes nothing, especially in a competitive place like Berkeley, where all faculty members were expected to make their way into the prestigious National Academy of Sciences, whose scientific gatekeepers were—impacters.

By the end of the decade, the battle seemed over. Scientists seemed to accept as fact that an impact had happened 65 million years ago and that it had caused the K-T mass extinction. And then, perhaps not surprisingly, a new prejudice arose. First in meetings near the end of some otherwise solid talk about yet another new K-T section, then at the ends of papers, the view that impact was a cause not only of the

K-T but also of others, or even all the mass extinctions, began to appear with regularity. If it could happen once, why not other times as well? Like some inexorable juggernaut, the Alvarez impact hypothesis somehow transformed into a larger entity: Unless shown otherwise, mass extinctions were always the result of asteroid or comet impact on Earth. This became the new paradigm, and it has held sway ever since. Till now, that is.

CHAPTER 2

The Overlooked Extinction

S outhern France, but not the southern France featured in books and movies. This place was far from Cannes or Provence, far from the Mediterranean, in fact. The blinding white quarry near the mineral bath resort town of Tercis-le-Bains was tucked up against the foothills of the western Pyrénées, the Bay of Biscay the nearest ocean. Its strata were therefore deposited in the same depositional basin as the previously studied K-T sites at Zumaya, Sopelana, Hendaye, and Bidart, places that had played such a large role in the K-T controversy in the late 1980s. But the Tercis site had accumulated its strata in very shallow water, compared to the deepwater sites on the Biscay coast, and consequently it held a wealth of shallow-water animal fossils.

The abandonment of the quarry, mined for decades for its pure white limestone, was a godsend to geologists. The last 10 million years of the Cretaceous period were found there. Unfortunately, while a K-T boundary was present, it had not been excavated at all by the quarriers and was covered by a riotous jungle of lush plant life.

Two geologists, festooned with the tools of their trade, stared at the quarry from the nearby bridge. They had just come from the small auberge near the quarry that so gracefully served them lunch each day, making this the most civilized kind of geologizing. It was an inn owned by a Socialist Party zealot, and since he gave a 10-franc discount to fellow socialists, the two geologists had become good socialists during their stay, even if they were somewhat mystified about what a good socialist did in France.

Each day had been spent in the Cretaceous-aged quarry, but today they decided to cross the river, where strata higher and thus younger from any in the quarry—younger, in fact, than any they had studied—could be found. Strata of Paleocene age. Here they could witness the aftermath of the K-T extinction.

A narrow footbridge led them across the slow-moving river that fronted the quarry, and soon they stood in front of a white wall. Fossils were everywhere. But the fossils were a disappointment to these veterans of Mesozoic digs, for gone were the iconic constituents of the Cretaceous period: There were none of the giant flat clams named *Inoceramus*, perhaps the most common animal fossil of the entire Cretaceous; gone too were the ammonites, also killed off by the K-T catastrophe. In their place were innumerable clam and snail fossils, and that was the rub. They looked identical to forms living today in the tropics. It was as if with the K-T extinction came an aftermath of modernization. These beds were Paleocene in age, and late Paleocene at that, strata that had formed no more than five million years after the K-T. And unlike the Permian extinction, where the first five million years of the Triassic remained barren of most life, these beds were vibrant evidence that animals had recovered quickly after the late Cretaceous impact.

There was not a lot of stratal thickness here, and they had no way of knowing as they approached the youngest beds here whether they were still in Paleocene rocks or were now in the overlying Eocene. But one thing was certain: In these top beds, the fossils seemed to come from an ever greater concentration of truly tropical species, including numerous corals that today are found only in the warmest oceans. It seemed that some long-term warming had gone on near the end of the Paleocene. It wasn't clear that this was of any consequence.

BY THE START OF 1990, A DECADE AFTER THE TWO-PART IMPACT HYPOTHesis was proposed by the Alvarez group, the field of paleontology in particular—and geology in general—was indeed unified in the belief that not only the K-T mass extinction but perhaps all of the major mass extinctions over the past 500 million years had been caused by asteroid impact. And why not? For in the early 1990s, the last bit of missing proof of impact, the crater itself, had been found. It was a gigantic, 120-mile-wide crater caused by the K-T asteroid, hiding in plain sight on the Yucatán Peninsula of Mexico. The crater was named Chicxulub after a small town nearby, and immediately after its discovery, plans were made to drill it.

It was a good time to be in the extinction racket, and now that the K-T case was closed, there remained lots of new places (and new old times) to research. To those in this line of research, it just seemed that all that they had to do was spread out over the four corners of Earth to unearth evidence of impact at every one of the major (and minor) mass extinction boundaries. Thus, a 1991 book by David Raup called *Extinction: Bad Genes or Bad Luck?* preached to an already convinced crowd, and his message filtered out to the general public, with eventual cinematic effect with the release of the certified Hollywood blockbusters of the latter 1990s, *Deep Impact* and *Armageddon*. In a concise style that eerily mimicked his speech (and perhaps thought processes too), Raup laid out in logical order his views on extinction in general and on why large-body impact was the prime cause not just of K-T but perhaps every extinction. The title of his tenth chapter summarized it all: "Could All Extinctions Be Caused by Impact?" Raup had his own agenda. In the late 1980s, along with his colleague Jack Sepkoski, he identified what he took as evidence of period extinctions. Every 26 million years, it seemed, a mass extinction occurred. But what could cause such periodicity? According to Raup and Sepkoski, if indeed periodicity existed (for it was based only on statistical findings, not on field evidence), there had to be some celestial rather than earthly cause. And into this breach stepped astronomer Rich Muller, who postulated that a faint and overlooked companion star to our Sun (which he named Nemesis) caused there to be heavier-than-normal asteroid flux crossing Earth's orbit every 26 million years. So with this background, Raup started his tenth chapter with the statement "Several times in the past couple of years, I have suggested to colleagues that meteorite impact might cause most extinctions." This sentence is but an amplification of the same statement made in 1988. In support of this hypothesis, Raup produced his now famous "kill curve," establishing the predicted number of extinctions to be expected from a range of asteroid sizes hitting Earth, from small to large. There remained only one small nagging worry: the strange "coincidence" between many of the major mass extinctions and the great flood basalt provinces scattered across the face of Earth, either on land or underwater.

Raup's book was the high-water mark of the view that most mass extinctions were caused by asteroid impact. This view was seconded in the introduction to the book, written by Stephen Jay Gould, and the one-two punch of the two biggest names in the field was considerable; it should be noted that since 1991 every one of the five biggest mass extinctions (as well as many of the lesser extinctions) has been linked to impact as cause in published literature.

While most paleontologists involved in extinction research during that time searched for evidence of impact, a different group, self-styled paleo-oceanographers, were retrieving data of a very different kind, data that while largely ignored soon after their publication, remain today explainable in only one way. And that way could have nothing to do with impact. Oddly enough, the information that would bring down one particular paradigm (that most or all mass extinctions were caused by asteroid impact with Earth) was discovered in the service of supporting a very different kind of paradigm—that Earth's contents drifted over the surface of Earth through a process called plate tectonics.

THE PALEO-OCEANOGRAPHERS WANTED TO SAMPLE SEDIMENTARY ROCKS from the deep ocean, and the series of ships dedicated to extracting drill cores from the ocean bottoms constructed two decades earlier, during the plate tectonics revolution of the 1960s, had launched a tour of the world, providing a means to do this. In the late 1980s, first the *Glomar Challenger* and then a designated replacement carried rotating crews of oceanographers and geologists to every corner of the seven seas, and in the 1980s the hubbub over the Cretaceous mass extinction was sufficient to stimulate a number of missions dedicated to extracting deep-sea cores with K-T boundaries enclosed. But of the entire oceans, the one place that the drill ships had difficulty in were the high latitudes of the Arctic and Antarctica, places where the brutally harsh weather conditions made drilling next to impossible. So another ship was obtained, the *Resolution*, and an eager team of scientists finally was able to penetrate the ocean bottoms near the poles.

A voyage called Leg 113 was to sail Antarctic waters, where a team that included renowned microfossil paleontologist Jim Kennett of Santa Barbara and Lowell Stott, then his student, began retrieving cores of strata with beautiful K-T boundaries contained within. But by the late 1980s there were fewer and fewer mysteries to be obtained from drilling yet another hole through the K-T boundary, and the phenomenal expense of this kind of work really stimulated the crew to try for something new.

And there was another motive, of course. For the more senior scientists there was always the goal of higher recognitions, with the pinnacle being elections into the National Academy of Sciences, a fixed-membership club that is one of the most prestigious and difficult to enter in the world. This was never spoken of, of course: Like the pilots in *The Right Stuff*, no scientist in his right mind would utter such an ambition. But it burns in most. And for the graduate students the holy grail was far different, and perhaps far more immediate and important. The American scientific community produces far more Ph.D.'s than there are jobs to employ them. Some number are never employed; of those that are, the vast majority hold government jobs or jobs at lesser universities where life is an endless succession of teaching class after class, and the smaller the school, generally the heavier the load. Such jobs never allow much time for research, and the grind soon burns out those who entered science not to be teachers but to do science, which is doing research. Only a very small number ever reach that holy place, an assistant professorship at a large and prestigious four-year university that promotes and facilitates original research. For Lowell Stott, that was the goal, and only a big-time discovery would enable him to reach it. Yet another K-T boundary description, while interesting (especially here at high latitude, to show how pervasive the effects of impact really were), would likely not seal the deal. But science is as much about luck as skill sometimes (although the old adage that luck favors the well prepared is also relevant here).

Kennett and Stott returned home with the samples from the trip, and two cores were especially interesting to them. The cores were named ODP 689 and ODP 690, the former being made up of strata deposited in quite deep water in the ancient ocean and the latter of strata from much shallower water. This was to turn out to be crucial.

While Kennett was trained to identify the microfossils that are found at the bottom of the seas, composed mainly of foraminifera (amoeba-like protozoa with a shell) as well as the smaller and plantlike coccolithophorids, whose skeletons when aggregated make up chalk, he had recently branched out into a new kind of research, the analysis of stable isotopes of carbon and oxygen that can be extracted from the tiny fossil shells in the cores. This kind of work, studying changes in carbon isotopes, was to turn out to be crucial in the study of mass extinctions and in differentiating one kind of extinction from another.

Carbon is not the only element whose isotopic differences can be used to great interpretive effect. Another is oxygen, with isotopes of ^{16}O and ^{18}O. As was the case with carbon, the lighter isotope, ^{16}O, is far more plentiful, and like carbon, the ratios of the two isotopes can be recovered from samples of clamshells or bone. The variance in the ratio of ^{16}O to ^{18}O has nothing to do with photosynthesis, however, but instead is related to the temperature of formation of the carbonate mineral trapping various oxygen molecules. In warmer settings, relatively less ^{18}O is taken up in the mineral, and with cooler environments the opposite occurs, in measurable fashion. This process has provided geoscientists with one of the most important of all tools, a virtual geothermometer.

Analysis of both carbon and oxygen isotopes was the goal of Kennett and Stott in analyzing the newly acquired Antarctic cores. The first samples run through the machines were from the K-T boundary parts of cores 689 and 690, and they yielded unsurprising results. Like those already found from other K-T boundary sites, the carbon isotope results from these deep-sea cores showed a simple pattern: They suggested that an extinction among photosynthesizing organisms had taken place. It was expected, because the reservoirs of carbon are linked in worldwide fashion.

But after having finished the K-T parts of the core (which did result in a paper in the journal *Nature*), they looked for something a bit more exciting, so the two kept analyzing their cores, moving upward in them and thus moving up through time. The cores represented a few million years of time, and in their upper reaches the two scientists found a second boundary.

This one was not a boundary between periods or even eras, as the K-T was. In fact, until that time it was regarded as a particularly boring interval in Earth history when little extinction took place and surely less likely to provide a scientific splash than even the K-T cores would. But being good scientists both, and having the chance to look for the first time at the carbon and oxygen isotopes of this time interval in a high-latitude setting, they went ahead and ran the samples through the mass spectrograph. What they found was at first a puzzle. Only later did they see it as a gift.

The numbers coming back from the oxygen isotope analyses at first glance suggested that some error in sampling or labeling had taken place. The oxygen isotopic values from core 689 were lighter than those from core 690 at the same time intervals. But core 689 was made up of sediment deposited in much deeper water than that of core 690. Even in the frigid Antarctic today, water cools with depth, and back in the surely much warmer Paleocene era, deeper water should be obviously colder than shallower. But the numbers here said just the opposite: warmer deep waters, cooler shallow waters. So they ran the samples again, with the same result. And this change was not seen in older parts of the core. Over a relatively short period of time, the deep ocean had anomalously warmed. Now *that* must have been a good moment, when the numbers first blinked from the machine, confirming this discovery. Luck does smile on the well prepared, and without intending to, Kennett and Stott had stumbled onto a major discovery with the only tools that could have made it.

Other evidence expanded the story. The carbon isotope record across the Paleocene–Eocene boundary in the two cores showed a short-lived "negative excursion"—the kind of record that occurs when the amount of plant life is reduced and so a hallmark of mass extinction. Other paleontologists began looking at the survival record of benthic organisms—bottom dwellers—from the region, looking specifically at the common benthic foraminifera, and found evidence of a catastrophic mass extinction on the bottom. This finding was even more sensational because these same creatures had suffered little in the then-recent K-T impact extinction. Was it simply that sudden warming of the deep wiped out the cold-adapted species in short order? That most pertinent question had not yet been answered in the early 1990s.

Kennett and Stott published their results in *Nature* in 1991. To a scientific community just then becoming accustomed to impact as a general cause of mass extinction, this discovery came as a mild shock—mild, because it was largely overlooked by those searching for impact-caused extinctions. But it was an important enough scientific discovery that did no harm to its two discoverers, and both received different but meaningful rewards that such a paradigm-changing find can give.

These discoveries showed that some 60 million years ago, the high-latitude deep ocean bottoms suddenly warmed, and as a consequence, benthic species died out. But this was no sensational K-T catastrophe. Benthic forams are not dinosaurs going suddenly extinct in fire and brimstone, and what at first seemed to be an event confined to the deep sea made no dent in a press mesmerized by rocks from space or in a coalition of geologists trying to make their bones by looking for the remains of other rocks from space at other mass-extinction boundaries. And there was more. In that same fateful year of 1991 a Japanese paleontologist named Kunio Kaiho published studies implying that the fate of the benthic forams was decided not only by rising temperature in the great depths of the sea but also by falling oxygen levels on the bottom. This made a lot of intuitive sense, for warm water can often become eutrophic and oxygen poor.

A deep-bottom warming, and lowering of bottom oxygen, even a warming of the surface waters. What was the ultimate cause? While some proposed the fashionable impact idea, this was quickly put into disfavor because of the pronounced differences in effects. The K-T event killed the plankton but left the deep relatively alone except for the loss of nutrients from above. Ultimately, it was correctly surmised that that entire warm bottom had come from the warm, tropical surface waters where evaporation would make the surface waters saltier and denser. This warm and saline water was then transported along the sea bottom like a conveyer belt, even as far as the cold, high-latitude sites of Paleocene age that were sampled by Kennett and Stott.

This was a sobering finding. It seemed as if the deepwater circulations of that long-ago time were different from what it is today, which sees conveyer currents running from the tropics to high latitudes, where cold, oxygen-rich water sinks to the bottom and flows back to the current's origin. This turned out to be quite correct, although the implications of this were overlooked at the time.

But just how catastrophic to the organisms of 60 million years ago was this Paleocene thermal event? Prior to the work of Kennett and Stott, the Paleocene–Eocene boundary had made no one's list of major mass extinctions; it hardly merited being a mass extinction at all, according to the compilation of extinction hunters David Raup and Jack Sepkoski. But part of that was omission, for at the time, little was known about the deep-sea foraminiferans, and with the discovery that so many of these benthic forams died out, and died out relatively quickly in an event that lasted about 400,000 years (but in which the switchover from one regime to another, the actual killing process, was much faster than this), the students of mass death began to pay more attention.

Still, to merit designation as a mass extinction at all, it would have to be shown that it was not just the ocean that was affected but land fauna as well. So beginning in the early 1990s, the search was on for any possible extinction on land that might have coincided with the vast number of extinctions (albeit among very small organisms) of Earth's ocean bottoms.

THE OCEANOGRAPHERS OF THE EARLY 1990S HAD DISCOVERED THAT A wholesale mass extinction among deepwater species had taken place at the end of the Paleocene epoch. Just the fact that the extinction took place among organisms (the deep-sea fauna) that had been about the only winners at the end of the Cretaceous period was certainly a curiosity. But did anything else on Earth feel this event, or was it entirely restricted to the deep sea? To do that, paleontologists needed to better understand the ranges of vertebrate fossils on various continents. The best place to study and collect vertebrate fossils of this age is in Wyoming's Bighorn Basin, and while nineteenth- and early-twentieth-century pioneers such as William Diller Matthew and Walter Granger had started the work there, it was a mid- to late-twentieth-century paleontologist, Phil Gingerich, who (with his students) studied and collected from these sections year in and year out, seminal work continuing to this day.

In the mid-1970s, the University of Michigan hired Gingerich, and from this base he set out to more fully explore the great biotic events of the time immediately after the demise of the dinosaurs. He convinced his colleagues to teach "field camp" (the capstone, field-work-based course that ends the coursework in every good geology department) in the Bighorn Basin and thus in one fell swoop had legions of newly minted geologists do the hard work of section measurement and collection in the huge area where these rocks crop out. Amazing amounts of material and information soon became available, and Gingerich was easily up to the task of publishing his results in a flood of refereed literature.

It did not hurt his cause that the field area lies in some of the most scenic country in the already scenic American West. From north of the Greybull River, and including the Beartooth and Absaroka ranges of the Rocky Mountain system, the thick piles of strata that had been deposited in rapidly subsiding river basins behind the rising Rocky Mountains were composed of sedimentary rocks accumulated by the numerous rivers, streams, and creeks with their attendant ponds, swamps, and small lakes that made up the area. This now-ancient place must have been alive with animals, big and small, and the sheer number of skeletons tells a tale of numerous corpses being carried by moving water to some quiet place, where bones were often rapidly buried by the rapidly accumulating silt, mud, and sand grain–sized particles eroding from the rapidly rising mountains. Sixty million years later, these graveyards are now thick piles of sedimentary rock, and it is within these that Gingerich and his students slowly, painstakingly, brought back the denizens of the Paleocene world.

Each fossil find had to be located with accuracy not only geographically, showing where it was found, but also stratigraphically, showing where in the thick piles of sedimentary rocks it came from. Species came and went, each showing some stratal thickness from the lowest (oldest) fossil find to the highest. By plotting the first and last occurrences, a table of ranges and occurrences was produced far exceeding in accuracy and detail anything that had come before. For just the critical time interval over which the greenhouse event took place in the Paleocene–Eocene transition, Gingerich and his students accumulated just slightly fewer than 20,000 individual fossil vertebrate specimens by 1998. From this they had a commanding view of what happened so long ago. It did not take long to see that a great turnover had in fact occurred among the mammals.

But was this turnover the same age as the marine turnover? Unless both sets of rocks (one deposited in the ocean, the other on land) providentially had volcanic ashes that could be dated using radioactivity half-life studies, there was no way of knowing if the newly discovered mass extinction on land happened at the same time as the mass extinction among the deep-sea organisms. To solve this problem, the various geologists involved in the project thought up a new way to compare the ages of the two groups of rocks being studied: They compared the pattern of carbon and oxygen isotopes. Sure enough, the patterns of carbon and oxygen isotopes showed that the mass extinction had happened at the same time on land as in the sea.

In terms of the fossil record on land, the event itself seemed to mark nothing less than the start of our modern mammalian fauna. While there were numerous kinds of mammals by the latter part of the Paleocene epoch (30 distinct families are recognized from the collected fossils), many of these were small, and some belonged to groups no longer present, including survivors of small and rodentlike forms, many kinds of marsupials, and some raccoonlike ungulates (a strange paradox, having the new entirely herbivorous ungulates taking on a meat-eating role in the Paleocene epoch). There were also true insectivores, the first primates (like the insectivores, still at small size). But by late Paleocene time there were larger forms as well, and some of these were truly bizarre. Dog- to bison-sized forms called pantodonts were leaf eaters that branched out into living a semi-aquatic lifestyle like that of hippos, or living in trees, as well as having larger forms moving about on all fours on the forest floors. In general they were stout of body, with short legs, and one cannot help but surmise that at least compared with modern herbivores, they were very clumsy and inelegant walkers. Yet large as they were, by the end of the Paleocene they were joined by even larger herbivores, the giant Dinocerata, looking like huge rhinos, even to the strange sets of knobs and horns on their skulls.

So all in all, any lucky time traveler to that ancient, warm, jungle-covered late Paleocene world could have filled any number of zoos with spectacularly different kinds of mammals, of shape and form guaranteed to stop the kids of our world cold in their tracks. Not dinosaurs, but pretty exotic and different creatures nevertheless. Little did they know, most were marked for extinction, or at least major evolutionary change. This world ended not with a bang but with a whimper—a hot one, at that—due to the distant happenings at sea and in the air.

Is there other evidence of what happened? It turns out that the amount of dust reaching the deep sea confirms suspicions that the world at the end of the Paleocene epoch became suddenly warmed, to the detriment of life. In the early and mid-1980s, investigators T.R. Janecek and D.K. Rea made clever use of deep-sea cores of the Paleocene–Eocene age to look at climate at that time. They reasoned that modern dust storms could provide a clue to the intensity of ancient wind systems. The desert areas where such storms occur produce what are termed Aeolian deposits, and these, it turned out, provided important clues to this problem.

Generally the rock record of these events is visible as sandstones with large-scale cross-bedding, such as the Navajo sandstone of the southwestern United States, an ancient desert dune field with nearby shallow-water beachfront turned to stone. But such deposits are rarely at best found in the deep sea; deserts do not have a habit of being inundated and carried down to the great ocean bottoms. But if the larger particles from sandy deserts do not get preserved as deep-sea deposits, the finer, dust-sized particles can. Large windstorms raising dust in an arid desert can produce sufficient velocity to raise smaller particles high into the stratosphere, where they are moved by the jet stream out to sea. These fine particles eventually fall from the sky onto the sea surface, where they slowly sink down onto even the deepest ocean depths.

In our world the amount of such dust making its way to the deep sea is a function of wind strength and the number of storms energetic enough to sending significant volumes of dust out to sea. If the number of such storms increases markedly, the amount of deepwater dust deposits increase in frequency and thickness, as well as the converse. But what causes such storms? Wind currents are ultimately related to the exchange of heat between pole and equator. When the average temperatures in these two areas of the globe are markedly different, the frequency and intensity of the storms will be greater. Janecek and Rea examined cores obtained from the deep-sea record to search for potential changes in this global temperature gradient by looking at the amount of dust hitting the Paleocene–Eocene oceans. They succeeded, but their results were a surprise.

Compared with today, the amount of dust making its way to the deep ocean over much of Paleocene time was rather less. But near the Paleocene–Eocene boundary they observed a striking threefold reduction in deep-sea dust. They also noticed another interesting rock type: volcanic ash. Like dust, this fine material makes its way to the seafloor from the atmosphere, but it is put up there by volcanic eruption, not atmospheric storms. While dust levels decreased across the boundary, volcanic ash levels increased. This increase could be due to only a sudden increase in global volcanic activity, about 58 million to 56 million years ago. Further work in many places around the globe confirmed these findings as being global phenomena, not anomalous events limited to one ocean basin.

Combined, these two records paint an interesting picture of that long-ago time. For reasons still poorly understood (but perhaps related to aspects of the ongoing opening of the Atlantic Ocean Basin), dear old Mother Earth blew her top with a fierce increase in volcanic activity, not just from land-based volcanoes but at the deep-sea spreading centers as well, as evidenced by an increase in hydrothermal deposits. At about the same time, the amount of wind dropped threefold on a globally averaged basis, and since the amount of dust is also a function of aridity, there is thus also evidence of global drying in a still unknown fraction of the continents. These latter changes could have come only from a reduction of the thermal gradient between equator and poles, and this interpretation is borne out by paleotemperature studies on fossils of that age. The late Paleocene tropics remained about the same (hot) temperature, but the Arctic and Antarctic regions warmed markedly.

Jim Zachos of Santa Cruz estimated that in a short interval of time the difference in temperatures from equator to pole changed markedly. Whereas in the Paleocene epoch the difference in seawater temperature between equator and pole was a hefty 17 degrees Celsius (it is an even heftier 45 degrees now), the difference had shrunk to only 6 degrees by early Eocene times. And as the high latitudes warmed, the heat exchange between the two regions slowed, reducing both the number and ferocity of storms. The world went calm and got very hot; a further consequence was mass extinction.

AT THE END OF IT ALL, THIS WORK ON THE PALEOCENE THERMAL EVENT, the various geologists, chemists, and paleontologists could distill a complex history ending in a lot of dead things on Earth down to an interesting series of events. First, Earth experienced a short-term rise in volcanism, and a consequence of this heightened number of explosive eruptions, as well as the more gentle eruptions, of flood basalts was a vast increase in the amount of carbon dioxide and other greenhouse gases entering the atmosphere. Very quickly, the air and oceans of our planet warmed. The warmer oceans contained less oxygen, and deep-ocean organisms, used to living in cool, highly oxygenated water, found themselves in the equivalent of hot poison and quickly died out. On land, the warmer air also changed the distribution of plants and trees, and some organisms died out. But it was a far cry from the wholesale destruction of the Earth-changing K-T event of only five million years prior. It is pretty safe to say that the extinctions in the oceans were more catastrophic than those on land. In neither case did the number of dead species come anywhere close to the extinctions in the Ordovician, Devonian, Permian, Triassic, or Cretaceous—the Big Five—each typified by the dying out of more than 50 percent of the species then on Earth. In the Paleocene event, the victims represented less than half this number.

How did the Paleocene event end? Eventually the aberrant rate of volcanism slowed. As the volcanoes let out less carbon dioxide, the upper atmosphere cooled, and eventually the oceans cooled as well. Was this a unique event? As the 1990s passed, more and more of the so-called minor extinctions seemed to show similarities to the Paleocene event. I was not to see one of these minor extinctions in the rock record until 2000, and by that time they were already thought not to be of impact origin.

THE TUNISIAN SOUTHERN DESERT, 2000

Donkeys. The Tunisian guide had been quite insistent that these donkeys were quite tractable beasts, the best sort of donkeys. (Of course they were saying this in French. Perhaps they were really saying that these were good Muslim donkeys and would not work for infidels.) We geologists, on this joint French-American expedition in July 2000 to sample late Cretaceous rocks in southern Tunisia, were quite uncertain how to deal with these familiar (through television) yet novel fieldwork tools. The donkeys had big baskets on their backs, plenty of room for the ungodly heavy suitcases that held the paleomag drills and attendant gear. Room, too, for the small gas tank. And most important, room for the huge carboys of water: water for the humans, water for the insatiable maw of the diamond-coated core drills, carried across one continent and one large ocean for the sole purpose of taking inch-long cores from ancient Cretaceous rock. To tell time, no less ancient time, for the purpose here was to better understand the age of the enormous white limestones that made up so much of southern Tunisia, and made up so much of northern Africa, in fact.

The rolling hills of white limestone, with cliffs so achingly blazing in the morning sun that even the teams' heavy-duty sunglasses let in way too much light for comfort, were a geologist's paradise, but the vegetation- and water-free environment was potentially deadly for humans. This was no place to break a leg or run out of water.

It was a long, two-hour drive for the team to get to this remote place from where they spent nights, in the town of El Kef, the last town as one heads south that could offer a clean hotel bed and reliable breakfast and dinner. Beyond that were only small villages that could have come from AD 1200, for electricity was just then coming to the south. The road south had one uncomfortable section, several miles where it ran along the border with Algeria before skirting back around toward the east. Algeria at that time was absolutely convulsed in mindless violence—the kidnapping and murder of Western oil specialists stupid enough or greedy enough to have not heeded the many U.S. State Department warnings to get out of that particular Dodge. The

Algerians had taken to border raids into neighboring Tunisia to punish what they viewed as a not sincere enough brand of Islam, favoring instead their brand, one that had seemingly come down from the Middle Ages. The punishment took the form of senseless slaughter, hand grenades, and small-arms fire into marketplaces, the wholesale murder of Tunisians. The small town nearest Algeria, where we geologists stocked up on water, had been hit hard the week before, Allah apparently not liking goods produced since the Industrial Revolution.

Some decades before, a now grizzled French geologist, Francis Robaszynski, had discovered the site they looked at now, and he was the guide on this trip, older now, more rumpled, dangling a cigarette in the French fashion. He had originally come here looking for the boundary between the Cretaceous and Tertiary periods, and he was not disappointed in what he found. This part of Tunisia indeed contains one of the best preserved K-T boundaries in the world, but the goal of the assembled scientists on that day was not to gather yet more information about that extinction but to attempt to coax a magnetostratigraphic record out of these deepwater limestones, one that could be matched with (and hopefully confirmed through replication) a similar magnetic stratigraphy worked out more than two decades earlier by Walter Alvarez and his colleagues, research that ended up with the great K-T discovery as well as a successful table of strata with its magnetic reversal pattern decoded.

Not too much farther south was the true Sahara. But this place of rock, vultures, and a few isolated huts among wadis seemed desertlike enough, and each morning the team would set out in the still-cool sunlight, the rocks still frigid from the plunging temperature of the desert night. On the third day here the route did not go directly to the highest Cretaceous but passed first through a series of meandering canyons of earlier Cretaceous age. The walls of the canyon grew progressively larger, and coming around yet another bend, the team was suddenly confronted by a gigantic wall of blinding white bedded rock, the remains of a shallow sea from about 100 million years ago, the middle Cretaceous period, when the changeover from saurischian long-necked dinosaurian herbivores to the rhinolike quadrupeds of the duckbill and ceratopsian lineages was taking place on land amid an even greater revolution, the great first flourishing of flowering plants displacing the long-dominant cone bearers in the world's forests, a change that opened the way for new insects amid this even newer gift to the world, flowers. But that was on land; the huge wall before us was from the sea, and it was packed not with land animals or plants but with the fossilized remains of marine mollusks of that time, small clams, snails, and the beautiful but now extinct ammonites, eventual victims of the K-T catastrophe that was still 30 million years in the future when these particular rocks were deposited on their quiet sea bottom. But interesting as the fossils were, they were not the reason that all save Robaszynski stared in amazement at the gigantic wall ahead. Halfway up its 100-foot height, a six-foot-thick layer of absolutely black rock was sandwiched like a negative Oreo filling, white outer coating gripping tight a black inner surprise, black as death, in fact, a true read of this layer's message.

The black layer was a proverbial sore thumb in these bleached white hills. Because the Tunisian hills, all made up of these layers of Cretaceous rocks, were sliced and slivered by faults of a long-ago tectonic paroxysm, the layers were not in the flat orientation of their origin on the sea bottom but were tilted, broken, bowed both upward and downward, the anticlines and synclines of Geology 101, and all of that structure was underscored by watching this so obvious black layer head outward across the hills into the far distance, and this was only its regional extent. The same black band can be found much farther than the eye can see, perhaps most spectacularly in the same aged rocks in the white foothills of Italy's Umbrian Apennines, where it is known as the Bonarelli bed, but also in the chalks of England, the shales of Wyoming, the thick limestones of Colorado, the offshore turbidites of California, the green siltstones of Alaska's Matanuska Valley, on an ammonite-covered island on the Queen Charlotte Islands off British Columbia. From place to place this black layer is variable in its thickness and even variable in shades of darkness, in some places just a thin stripe of brown but everywhere the same message: death and mass extinction, the world strangling in the so-called Cenomanian–Turonian event.

I was very startled by this black bed. It was not a case where the limestones became increasingly dark approaching this bed. No, this was a day-and-night change. I came closer, touched it, smacked it with my hammer, sending shards of hard limestone in all directions, including that of our seated chief, Robaszynski, who was rolling another cigarette between yellowed fingers when the limey shrapnel came his way. I searched for any indication that the black bed sat on the older beds unconformably—the term we use when erosion has removed rock, so that any new deposition makes the first beds look like they are part of a continuous deposition of strata, rather then episodic deposition. If this had happened, some underwater current could have removed rock, showing a gradational change between the white, oxygenated rock, and this black rock that came into being on a sea bottom with little or no oxygen. But no. It was like a switch had been thrown: white limestone, filled with fossils, an indication of a living sea bottom. Then flip, the switch thrown, and Black Death is the record, as oxygen-free water rather quickly asphyxiated the previous inhabitants of this bottom.

The black band of sedimentary rock that confronted that particular team so starkly on that day in Tunisia has been known to geologists for a long time, but it was not until 1970 that its origin began to be scientifically probed. In that year Seymour Schlanger and Hugh

Jenkyns began compiling a list of localities where this strange black band could be observed. Further work showed that it was not a single event but three events spaced out over about 20 million years when the entire ocean went stagnant—currents stopped and the deep began to suffocate as organic matter raining from the plankton above settled onto the bottom and began devouring the oxygen stored there, with no oxygen coming down to replace it. And not content on just killing the bottom dwellers, this now enlarging body of low- or zero-oxygen water grew upward like some B-movie blob. Sediments deposited first during the Albian stage, 110 million years ago, and then at the end of the Cenomanian stage show all the hallmarks of anoxia: They are finely bedded, because the normal bottom burrowers such as worms, sea cucumbers, and crustaceans were killed off, letting the fine layers accumulate. There is even a pattern to the beds that can be related to changes in Earth's orbit first identified by Russian geologist Milutin Milankovich, cycles now bearing his name that are linked to the current ice ages as well as the thickness and spacing of these deep sea limestones and shales.

Schlanger and Jenkyns did more than observe these black beds. They used the then fairly novel method of carbon and oxygen isotope analysis. They found that the microscopic shell-builders that had lived either in the surface water or on the deep bottom had the same carbon isotope signal. This pattern would prove essential to the later work on the Paleocene deep seas—showing that a change in the temperature of the ocean had obliterated not only individuals but also species.

SEVERAL MASS EXTINCTIONS HAD NOW BEEN TIDILY EXPLAINED, ALBEIT they were pretty minor ones. They were caused not by asteroid attacks but by warm oceans. Trouble was, nobody listened—after all, these were only minor extinctions. And with the end of this work in the mid-1990s, very little new work on extinctions continued, other than more mop-up from the K-T and planning a series of expensive cores of the Chicxulub crater. Although this was a quiet period, new methods and methodologies were being developed in the wings, and campaigns were being planned to attack two of the Big Five: the Permian and Triassic mass extinctions, which, in the 1990s, were certainly thought to have been caused by gigantic asteroid impact with Earth. And sure enough, early in the new century, new evidence from both the Permian and Triassic mass extinctions would reinforce that opinion.

CHAPTER 3

The Mother of All Extinctions

BETHULIE, ORANGE FREE STATE, SOUTH AFRICA, OCTOBER 1999

I was tired, bored, hot, thirsty, and very much wanting to go back to camp and rest my sore feet in a bowl of muddy-but-cool river water. With skin pinked by the sun and wrinkled by the incessant, sandy wind, I shuffled through the dusty heat on sore legs, arriving at a fork in the small watercourse that I had been following for more than an hour, vainly seeking paleontological gold, the spectral skulls and skeletons of the long—the very long—dead. Only patches of sedimentary rock were visible above this sandy streambed, one that eventually emptied into the Orange Free State's Caledon River, itself one of the largest watercourses in this dry region of South Africa's Great Karoo Desert. One fork circled roughly back in the downhill direction I had come from, leading downward through time, back into the Permian period, toward the slightly older rocks at the river's edge that were full of Permian-aged skeletons, the remains of a large and curious land animal fauna that characterized planet Earth some 251 million years ago. The other fork headed in the opposite temporal and geographic direction—up in time toward the Triassic period, the time immediately after the greatest mass extinction in Earth's history, the Permian extinction, also known as the Great Dying or the Mother of All Mass Extinctions.

So who was the father? I mused mirthlessly, deciding to keep heading up through time, even if it took me farther from the truck and my companions. This was no small decision, especially this late in the day. While fossils had been common on the Caledon, in beds deposited perhaps a million years before the Permian extinction, they had become ever rarer as I approached the end of the Permian. This was so different from my earlier experience in studying mass extinction, for I had come here following a decade of studying the fossil record at many K-T boundaries around the world, and at every one of those places the fossil record had been very different from here indeed. The Cretaceous fossils remained common and diverse right up to the K-T impact layer with its overlying boundary clay—and there they simply disappeared. Here in these late Permian rocks, vastly older than even the Cretaceous, it was as if the world had been slowly dying over a considerable length of time. There were many possibilities for this: Perhaps the nature of the way in which these large land animals had died and become entrained in sediment, to ultimately fossilize and rest a quarter of a billion years awaiting disinterment, had changed. Perhaps rivers had dried up, and the subsequent traps for bones disappeared as well. But perhaps the animals themselves gradually became rare, as some longer-term hand slowly but inexorably closed the windpipe of a living Earth to a near-death experience.

A particularly eye-loving fly ended that reverie, and in spite of the near absence of fossil material, I began to search again for bone, any bone, jutting from the olive-colored sedimentary rocks, striving to become an automaton, a living machine bent on seeing the visual cues of ancient bone: colors, textures, shapes that would subtly call to the

THE MOTHER OF ALL EXTINCTIONS

prepared eye and mind. I looked up for bearings: Large buff cliffs a half mile distant were made up of sandstones and red beds of Triassic age, while the greenish rocks that peeked out of each twist in the watercourse were definitely Permian, but I was damned if directions made any sense here in the southern hemisphere; the only thing that was dependable was that the sun set back over the river. If that was Triassic up there, then somewhere ahead, and not far, had to be rock that had accumulated in swamps, lakes, ponds, but mostly in river valleys during the time of that long-ago cataclysm.

I got dustier step after step and took an occasional swig of water from my diminishing supply, sweat coursing out of my skin to be immediately swallowed by the dry air. My increased concentration was almost immediately rewarded: a few broken, eroded, but unmistakable fossils of the last Permian animals, all preserved not as articulated skeletons as they were on the Caledon's wide stratal riverbanks but as isolated bones and teeth. Here a large scapula, probably from the most common and characteristic animal of the latest Permian, the large, cowlike *Dicynodon*; there a tusk of another *Dicynodon*; and most sensational, the broken tooth of the most fearsome carnivore of that long-ago world, a gorgonopsian, or Gorgon, as the paleontologists called them. *Permian, Permian, Permian*, the rocks whispered to my increasingly addled mind, the heat probably insignificant compared with that at the end of the Permian period but hot enough to make a human brain continually wander, to lose focus and capability.

Eyes down on the ground as more and more outcropping began to appear along both sides of the creek bed, I rounded a corner of the increasingly higher walled gulley and nearly died of a heart attack as three elk-sized, long-horned gemsboks startled into flight, leaping upward to scramble out of the gully, flailing legs scratching a shower of pebbles and soil out of the dirt walls as they struggled in panic to run from this bipedal stranger, spraying him with sediment in the process.

Heart beating stoutly, I sat down huffing, swinging the heavy pack over my shoulder onto the ground, and rummaged for food of some sort to calm myself a bit. The short burst of adrenaline spurred by the game was dissipating, lassitude returning, only an hour till pickup and camp, an hour until that delicious first beer capped a hot day of fossil collecting. An hour. What to do for yet another hour? Idly looking down at the rocks I sat on, I absentmindedly watched the purposeful ants marching to and fro before I focused on the ants' freeway. The ant-covered sedimentary rocks were strangely colored compared with the strata that had been present all afternoon. Not the drab olive of the Permian or the bright red of the Triassic just ahead, but an anomalous candy-cane assemblage of both.

Curious now, I stood and followed the strata to the gulley wall, to be immediately confronted by a beautifully clean rock surface, obviously scoured annually by the occasional flash flood that the Karoo experiences in its June-through-September winter. If anything, the thinly striped alternation of red and olive was even more pronounced here, about a half inch of each, beds that were clearly laminated. Such beds are known to be preserved only in the absence of life. Actually, all the day's beds originally were just like these, but soon after their formation, ancient, Permian armies of insects, nymphs, worms, crustaceans, even the shuffling feet of the larger vertebrates visiting the shallow ponds and waterholes, where the sediment was accumulating destroyed the fine laminar bedding, churning it into a mass mixed mud of one color and almost devoid of any layering at all.

Excited now, I climbed upward, for the elevation was rising toward the high sandstone hills in front of me, and my climbing took me into younger parts of the flat-lying sedimentary strata. Only a few tens of feet above the striped rocks, themselves at least a dozen feet thick, the characteristic red mudstones with their small white limestone nodules characteristic of the Triassic were brazenly visible, and within a few minutes, an eroded skull of a small *Lystrosaurus* confirmed the suspicion that this was the lowermost Triassic. I walked a few tens of feet higher into the Triassic strata and entered a fossil hunter's heaven: first tens, then hundreds, of bones, showing as a characteristic and easily seen blue-white color amid the red rocks, all from the pig-sized mammal-like reptile, *Lystrosaurus*, the index fossil that characterizes the earliest times after the Permian mass extinction. After the long days of searching the uppermost Permian, with its beds so barren of fossils, it was sheer joy to be amid such treasure. But these fossils told little not already known, and they were all of but a single species, rather than the more than 50 species in the Permian beds below. But it was a temptation to stay here and collect ever more wondrous fossil treasures just for the joy of it.

Discipline kicked in once again—turn around, return down into the creek, arriving in minutes once again at the striped beds. There I became busy with camera and notebook, recording thoughts, observations, and measurements. The sun had dropped into its late-afternoon position, lighting the striped strata into ever-brighter relief. I had never seen rocks like this lower in the Permian nor higher in the Triassic, but memory leafed through its files, and there were vaguely remembered beds like this at the sites called Lootsberg Pass and nearby Wapadsburg. But at those two places, this part of the sedimentary transition had been highly weathered, and since the job then and now was more about finding fossils than about noting the nature of the rocks, I had thought nothing of it. But here these rocks were cleaned, presented in their best aspect, daring me to ignore them. This creek itself was a new discovery, and its trove of fossils spoke of my being the first paleontologist to ever collect it, in all probability. So no geologist had ever stood here, to see what would soon be recognized as the best-exposed Permian–Triassic boundary in the vast Karoo Desert.

But what caused the thinly bedded rocks to form, sandwiched as they were between the normal-looking Permian rocks below and Triassic rocks above? I now knew the "stripy" beds were near what had to be the P-T boundary, and, hand lens gathered in front of face, nose to rock, I began to scan for the thin impact layer that I had so often seen in the younger Cretaceous-aged rocks, and the impact layer I absolutely believed to be present here as well. Here in 1999, I still hoped to be the first to prove my—and others'—hope for the Permian, with evidence of a large asteroid impact, evidence that would show that the Permian extinction, just like the Cretaceous mass extinction, was caused by impact as most of the geology community believed—and had believed since 1990. I wanted to be the Permian version of Walter Alvarez, and if that led to a pleasant basking in the publicity and academic honors that would come to the first to truly show that the greatest extinction was an impact extinction, so be it. I could learn the necessary modesty while pocketing the pay raise. The only nagging problem was that no matter how I and so many colleagues all over the world tried, no one had been able to find any of the telltale evidence of impact so clear and abundant in both marine and land deposits of 65 million years ago, the day that the dinosaurs were very quickly killed off. That was a round hole of evidence, and all that could be found in the Karoo and the many other P-T boundary sections were the square boxes that we tried to smash our meager evidence into.

I scanned the surface with my lens again, found nothing, and patiently began the process again. Small samples were collected across the many strata making up these beds. The sound of a distant truck horn signaled the end of the day. But I would be back, dragging my companions the next morning. This place was important. It just had to be impact. How could slow climate change, or volcanoes with lava thick or thin, have caused this greatest mass extinction in Earth's history?

WITH THE END OF THE TWENTIETH CENTURY AND THE ARRIVAL OF THE twenty-first, ever more attention was being paid to the Permian extinction, and why not—it was the largest of all, with as many as 90 percent of all species disappearing. But how fast, which is a clue to how, began to be best appreciated with the work of paleontologists from China and the United States in extensive studies of thick Permian and Triassic limestones cropping out near Meishan, China. The geologists in China even had an advantage not present at most of the K-T sites—in China there were scattered ash layers that could be dated using large machines, and this was done on samples by MIT's Sam Bowring. The results of this vast enterprise came out in 2001 in *Science*, authored by Y. Jin, Doug Erwin, Sam Bowring, and other Chinese colleagues.

The China effort combined results from five different stratigraphic sections in the Meishan locality, with sampling intervals made every 30 to 50 centimeters. A total of 333 species of marine life were ultimately found in these rocks, belonging to such varied sea creatures as corals, bivalve and brachiopod shellfish, snails, cephalopods, and trilobites, among others. Nowhere at any stratigraphic horizon at any time has so thorough a collecting effort—or so rich a fauna—been documented with such precision.

The authors did indeed find one horizon where more fossils went extinct than in any other, in the last meter of strata deposited at the very end of the Permian period, and this was like the situation at the K-T boundary sites. But unlike the K-T sites, which showed but a single level of mass extinction coincident with the impact layer, these Permian sites showed many other levels in which lesser but still significant numbers of fossils suddenly went extinct in addition to this most catastrophic level. These layers both predated and postdated the P-T boundary, which was identified based on the last occurrence of small fossils known as conodonts. It was as if there had been a series of catastrophes, one big one and many nearly as big ones.

Some years earlier, one of the coauthors, Erwin of the Smithsonian, had championed a theory that he called the *Murder on the Orient Express* explanation: that just as there was no single killer in the great Agatha Christie whodunit, so too was the P-T event really the end result of Earth undergoing a multitude of stresses that when combined, caused the hideous mass extinction. But in 2000 Erwin came to a new view. His work with Sam Bowring on the Chinese sections had shown that the event must have taken place in 165,000 years or less, with emphasis on the "or less." But this is still a far cry from the interval of time that was by that point accepted for the K-T die-off—not hundreds of thousands of years but perhaps just decades.

Like so many others consumed by the mysteries of mass extinctions, Erwin searched for cause among the many threads of evidence left behind in the rock record. The various environmental conditions in the seas at the end of the Permian included widespread evidence of oceanic anoxia, or low oxygenation of seawater, in both the shallow and deep sea. The anoxia was apparently of such magnitude that many marine organisms were rather suddenly killed off, just as they are today in modern red tides. There is also evidence of global warming at the time of the extinction, and the coincidence—if that is what it was—of the Siberian lava eruptions at the same time as the mass extinction. And—the elephant in the room—there had even been the sensational, mid-1980s announcement from a Chinese group that they had discovered an iridium-rich impact layer from the highest Permian rock in these fossiliferous sections.

But science is predicated on replicability. American researchers asked for splits of the Chinese samples, and to the ultimate embarrassment of the Chinese, the highly sensitive American instruments could find no hint of excess iridium. When the dust settled, there was no indication of impact from these rocks.

How to account for the various lines of evidence that did hold up, and how could they add up to a possible, single cause—if at all? Erwin summarized the various suspects. First is the possibility that the Siberian traps introduced large volumes of gas into the atmosphere, triggering large-scale climate change and acid rain, as earlier suggested by Paul Renne and others. With new information from disparate sources, a sudden methane release into the atmosphere became a viable candidate for the killer. But in spite of no evidence to support impact, the understanding that impact could cause extinction was still on everyone's mind. The new evidence from China argued for some sort of "quick strike." Among potential causes of mass extinction, only asteroid impact was thought to be capable of causing such mass death in so short a time. The last sentence in the report by Jin et al. says it all:

> Despite the lack of compelling evidence for extraterrestrial impact, the rapidity of the extinction and the associated environmental changes are also consistent with the involvement of a bolide impact in this most severe biotic crisis in the history of life.

Thus, in 2000, the Permian extinction looked like nothing known— it was still suspected to be some sort of impact extinction by the geological community, but one seemingly different from the K-T event: perhaps many impacts or a single large impact superimposed on some other kind of extinction mechanism. The most puzzling thing was that search as they might, none of the investigators looking at the Chinese rocks could find the well-known clues so common at the many K-T boundary sites. And then, as if the geological gods had answered the prayers of geologists beseeching them for impact evidence at the end of the Permian, in one fell swoop new results from three Permian outcrops, including the crucial one at Meishan, China, fingered impact as the culprit after all. This new evidence came from an entirely new line of geochemical study. For those yearning to find impact at all mass-extinction boundaries, a strange substance known as buckyballs seemed to come to the rescue. But in fact, what they did was light an ongoing controversy.

It was in 2001 that a new character emerged center stage with a dramatic report published in *Science*. The senior author was a geochemist trained at the Scripps Institution of Oceanography named Luann Becker. Her colleagues were Robert Poreda and Andrew Hunt from the University of Rochester, New York; Ted Bunch of the National Aeronautics and Space Administration's (NASA's) Ames Research Center at Moffett Field, California; and Michael Rampino of New York University and the Goddard Institute for Space Studies. They reported finding, in the critical latest Permian boundary layers, high levels of complex carbon molecules called buckminsterfullerenes, or buckyballs for short, with the noble (or chemically nonreactive) gases helium and argon trapped inside their cage structures. Fullerenes, which contain at least 60 carbon atoms and have a structure resembling a soccer ball or a geodesic dome, are named for Buckminster Fuller, who invented the geodesic dome.

The researchers interpreted these particular buckyballs as extraterrestrial in origin, and therefore, like iridium (which, pointedly, was *not* found) because the noble gases trapped inside have an unusual ratio of isotopes. For instance, terrestrial helium is mostly helium-2 and contains only a small amount of helium-3, whereas extraterrestrial helium—the kind found in these fullerenes—is mostly helium-3. According to the authors, all this star stuff could only have been brought to Earth by a comet impacting Earth at the end of the Permian period (more correctly, it ended the Permian). They found this stuff by sampling as if for carbon isotopes—by taking lots of bits of rock both above and below the boundary, carrying the pesky things though U.S. customs in the cases, and then analyzing them back in the United States.

Back in the United States, the Becker team broke down its rocks in search of the buckyballs. They are not visible to the naked eye and can be confirmed as present only by using a special kind of mass spectrometer. The results were spectacular, as results from each of the samples sections in China and Japan (but not at the third site, in Hungary) flashed onto the computer monitors attached to the various mass specs. According to the authors, the Chinese and Japanese samples were striking in being packed with evidence that Earth had been slammed by a comet (or asteroid) at the end of the Permian period. Fullerenes were found at very low concentrations above and below the boundary layer at the two sites, but they were found in unusually high concentrations at the time of the extinction.

Not only was an impact confirmed, according to the team, but also the quantitatively determined mass of buckyballs even allowed them, the authors said, to estimate the size of the comet. The researchers announced that the comet or asteroid was 6 to 12 kilometers across, or about the size of the K-T asteroid that left the huge Chicxulub crater. The scientists had arrived at this size estimate on the basis of two factors—if the body were smaller than 6 kilometers in diameter, the effects wouldn't be seen globally, as they appear to have been; and if it were larger than 12 kilometers in diameter, there would have to be more gas-laden fullerenes distributed globally.

No one likes to be scooped. By this time, a lot of scientists had been looking for evidence of impact at P-T boundary sites for years without success. Out of the blue, a new team had hit pay dirt, and much was at stake: research money, professional advancement, but most of all, pride. Scientists are human. Of course there were very sour grapes. The results from Becker et al. were intensely scrutinized by the many who had searched without success for impact evidence at the end of the Permian period, each work weighed, each number pondered, each conclusion considered. Not surprisingly (for scientists are a naturally skeptical subspecies of humans), doubts arose, e-mails flew, and long-distance telephone charges grew.

It was the estimation of the impacter size that first made a number of the mass-extinction clan suspicious of the whole thing. Later, other doubts arose, but the impacter size jumped out at many K-T veterans when the first draft of the manuscript by Becker et al. was sent to colleagues for prepublication scrutiny. Not because the impacter size was not appropriate—on the contrary, the size was the same as that causing the later K-T catastrophe. Doubts arose because the estimated size was too perfect.

Even before publication, those asked by the Becker group to helpfully vet the manuscript were pretty much brushed aside. The publication appeared, the press had a field day, and Luann Becker thus first appeared on the scientific and public stage in dramatic fashion. She was no novice in science—prior to the Permian buckyball paper, she had published a number of papers about meteorites and their chemical compositions, and, for instance, her work on the chemical composition of some well-known meteorites in collections had nicely increased understanding of the chemical compositions of some of these widely varying kinds of rocks from space.

But it is safe to say that until 2001, she was but another of the army of scientists trying to figure out the chemistry of the cosmos. But that relative anonymity utterly disappeared with the buckyball article in *Science*, and soon word of this discovery was trumpeted by its major funder, NASA, in a large press conference held in Washington, D.C. NASA even invited the most experienced of all Permian workers, Doug Erwin, and he was surely bemused by all this sudden attention on a problem that he had rather anonymously worked on for years. All of a sudden the Permian extinction was one of the hottest scientific issues going—impact had once again done its magic with the press.

It would seem to be hard to follow up the 2001 circus about bucky-balls and the Permian extinction with anything as dramatic, but two years later the Becker crew did just that, using the same sequence of discoveries that had characterized the history of K-T research: In 2003 they announced, again in *Science*, that they had found the crater of the P-T impacter itself, the source of all of those buckyballs and helium-3. Once again the whole issue made big news, the reports of two years earlier that the Permian extinction was now "proven" to have been caused by an impact quite forgotten. And this was not the only candidate "crater" to be found. It mattered not that the most experienced student of impact craters and their origin anywhere in the Solar System wrote that the structure identified as the Permian impact crater by the Becker team looked like no other impact crater in the Solar System. (Simply look at our Moon or Mars or Mercury to get a sense of how many impact craters there are in our cosmic neighborhood. To be unique among such a large number is pretty unlikely—or, the structure in question is not an impact crater).

And this was not the end of things. In 2006, a team from Ohio State University announced (at a scientific conference, not in print) that a large structure far beneath Antarctic ice was probably "the" Permian crater—but this was really bad science, since they could neither confirm that the large structure, detected remotely using gravity anomaly measurements (large craters give a different gravity reading than surrounding rock), and most important, since they could not reach any of the buried rock, they had no way of knowing what age their "crater" was. Nevertheless, once again the press grandly announced that the Permian extinction mystery was solved and that it was caused by impact.

Back in 2001, Becker and crew probably expected to be met with open arms and praise by the many scientists who had shown that the K-T event was caused by impact. But there were not a few cold shoulders. Some of this might have been jealousy, for who does not love the attention of the fickle press, especially scientists trying to get funded. Yet there was more to this doubt than that. There was a nagging unease about the data. The whole issue of buckyballs and helium-3 had yet to be accepted or even, most important, replicated by other labs. There was also a distinct feeling that not everything written by Becker et al. added up scientifically. For instance, one of the lead paragraphs of their first press release stated: "The collision wasn't directly responsible for the extinction but rather triggered a series of events, such as massive volcanism and changes in ocean oxygen, sea level, and climate." This conclusion made no sense at all.

How could a comet impact create volcanism or a change in sea level? Much was known about what large-body impact on Earth could or could not do, and this was in the realm of the "could not do." While it makes intuitive sense that a large rock slamming into Earth could somehow shake free some great volcanic paroxysm, that doesn't mean that it will. About this time many workers became less sanguine about the possibility that the P-T extinction had been caused by impact, leaving buckyballs and dead species in its wake.

The first meeting of what was to become a loyal (to science) opposition was convened only several days after the initial publication by Becker et al. in 2001, and it was by sheer coincidence that an eminent group of specialists most concerned with the P-T came together. Such a gathering was sure to eventually take place, but it might have been a year afterward or more, perhaps much more, at some scientific meeting or the other, that these same scientists would have most certainly compared notes on the purported Becker discovery, if discovery it indeed was. For other reasons than talking about the P-T extinction, Doug Erwin of the Smithsonian and Yukio Isozaki of Japan arrived simultaneously at the Division of Geological and Planetary Sciences of the California Institute of Technology in Pasadena. Already there were two California Institute of Technology (Cal Tech) faculty members also immersed in P-T research, Joe Kirschvink and Ken Farley. They decided to spend an afternoon going over the *Science* paper by Becker et al.

These four sat around a table, chewing on the Becker paper, and were immersed in the give-and-take of critical science. Of the four there, Farley was by training the most versed in the primary argument of the Becker report—that helium isotopic ratios could provide evidence of a past asteroid or comet collision with Earth. In addition, Farley was already acquainted with Becker, as both had been grad students at Scripps Institution of Oceanography, in La Jolla, California. Following his thesis research, Farley had gone on to do work on noble gases, and in fact by that time he was recognized as the world's authority on helium in rocks. Among the others around the table, Erwin was a macrofossil paleontologist and the acknowledged expert on the extinction of larger fossils at the end of the Permian period, while Isozaki nicely complemented the paleontological side of things, as his specialty was the identity and fates of microfossils before, during, and after the mass extinction. Kirschvink, the last member of this group, had by this time spent several years in the Karoo of South Africa looking at the P-T boundary. It would have been hard to find a better group for critical analysis of the data from Becker et al.

As the first step in this process, the small group studied the data tables in the paper by Becker et al., fixing on the amount of buckyballs found at the three sites. While Becker had touted Hungary as yielding the crucial carbon compounds, once the four dug deeper into the data part of the article, they concluded that the Hungary site showed *no* evidence of fullerenes, so the critical evidence came from the other two sites. Both of these suites of rocks—from China and Japan—were intimately familiar to the assembled group at Cal Tech, for Erwin and his colleague Bowring had collected the Chinese samples analyzed by Becker, while Isozaki had done the seminal work on the Japanese P-T boundary that Becker and her group had analyzed. And it was here that Isosaki let loose his bombshell, still unknown to Becker.

Geologist Rampino, an absolutely die-hard proponent of impact and one of the authors of the *Science* paper, had collected the samples from Japan from seriously deformed and highly fractured deep-sea sediments. Isozaki, a specialist on deep-sea microfossils of this age, had gone back, following sampling by Rampino but prior to the publication by the Becker group, and looked at the site of the crucial samples. The places sampled were obvious, and if they were of latest Permian age, they would contain a specific set of latest Permian microsamples. To his surprise, the fossils found by Isozaki from these rocks were not the Permian species at all but were Triassic in age, from a time when absolutely no extinctions took place! Unknown to himself or any other members of the Becker team, Rampino had sampled rocks far *younger* than the crucial Permian age! Any results had nothing to do with the Permian mass extinction.

Thus, with no buckyballs from the Hungarian samples and the discredit of the Japanese samples, that left only the samples from China as proof of an impact. And it also left a huge residue of unease among the P-T specialists. Kirschvink, Erwin, and Isozaki turned to Farley, for he alone would be able to repeat Luann's observations on the critical Chinese P-T samples. Farley had by that time come up with a less laborious way of detecting helium-3 from rocks. After all, it was not the buckyballs that were so important (for they can be made on Earth, in forest fires, for instance) but the fact that they encased the helium-3. Farley bypassed the buckyball question entirely, going straight after the amount of helium in the rocks. He even tested his new method— on K-T samples that Kirschvink and I had obtained in Tunisia the year before—and had found helium-3 in them, as was expected. Now, at the urging of all, Farley turned his attention to the Chinese samples.

All science is predicated on replicability, as belabored earlier in this chapter for a reason. Farley contacted Becker, asking for splits of her critical Chinese samples. She replied that she had used up all of her samples in the analysis and could supply no more. This was curious and unsettling—how could she have *not* saved some of the samples so that others could do just what Farley was trying to do—replicate her results? Farley then went to the original source of the samples, MIT geologist Bowring, who had actually taken the critical samples from the Meishan locality that was now so crucial for understanding the ancient mass extinctions. Bowring promptly sent new material from China, which was duly analyzed. Farley used a blind sampling technique, asking Bowring to withhold any information about which samples came from the critical level where Becker had found the helium-bearing buckyballs. After exhaustive tests, Farley was not able to replicate the Becker group's findings. There was no helium-3 to be found in the Chinese samples examined by the Cal Tech lab specializing in this kind of work.

Speculations were rampant after this surprising development. Becker was told of these results and shrugged them off; the most likely reason for this negative finding, she reasoned, was that the helium-3 layer discovered by her group was an extremely thin layer from a more massive sample collected and supplied by Bowring and that the material later sent to the Cal Tech group did not contain this exact bit of rock. But others came to other conclusions. There was some speculation that the Becker findings might have somehow been related to lab error or contamination of Becker's glasswear in some fashion.

Becker pressed on, of course, and as recounted above, continued in the lines of K-T science by following up the geochemical discoveries with an announcement that the team had found the crater left behind by the comet spewing all those buckyballs over the planet (but by-passing Hungary, anyway) in the paper published in 2003. The Bedout crater, as it was named, lay underwater off Australia and was certainly large enough, containing rocks from within that were of approximately the right age. Another rock from space (or ice ball, in this case) did the damage. Impacts cause extinctions, a paradigm again verified.

Soon thereafter, letters bombarded *Science*, demanding to know how such work could get published, and *Science* went silently defensive (probably to the delight of the competing European journal *Nature*, for some of the harshest critics of the whole Permian impact story were European impact specialists). But as far as public relations went, who cared? It was a good story, tidily completed.

Thus, by the middle of the first decade of the new century, the riddle of the cause of the Permian extinction was solved, at least in the press's and public's minds, by the discovery of nonreplicable helium-3 findings from a noncrater crater. What a disconnect between the public and the on-the-ground scientists!

So if not impact—what? In the first five years of the new century, two camps emerged, each deeply entrenched in its views: Either the Permian extinction was caused by impact or its cause was unknown but certainly *not* impact. In favor of the former was the discovery by the Jin and Erwin group of a sudden extinction exhibited by the fossil invertebrates. How else but impact could this have occurred? Yet many lines of evidence were converging on something *more* prolonged than a single quick strike. In September 2000, University of Oregon geologists Evelyn Krull and Greg Retallack published a paper detailing their results from prolonged geological and geochemical studies of P-T boundary sections in Antarctica. Their results strongly supported the idea that the early Triassic was a time of heightened methane gas volumes in the atmosphere. Methane is one of the most potent of the greenhouse gases—and its sudden release would have driven global temperatures sharply higher. These results followed on Retallack's 1999 findings from the Sydney Basin in Australia. There, Retallack recognized that the P-T boundary was coincident with the formation of the last coals anywhere on Earth for many millions of years of Triassic time.

The boundary coincided with a large-scale extinction among plant species as well as a dramatic changeover in climate, as deduced from fossil flora and fossil soils. A deciduous flora adapted for a humid but cold temperate climate characterized the latest Permian of Australia. At that time, Australia, like nearby South Africa, was located far nearer the poles than the equator. In the earliest Triassic, however, a marked change in climate apparently occurred. The fossil soil types indicate a much warmer climate—as would occur from a sudden onset of global warming. Coal formation suddenly ceased. Sedimentation rates markedly increased in the lower Triassic rocks, and Retallack interpreted this as being the result of extensive and sudden deforestation at the P-T boundary.

Other suggestions of a profound world-changing event came from Roger Buick, a geoscientist from Australia. Buick, a specialist on the Precambrian world, became intrigued with the P-T event because of how it sent our world, for a short time, back to conditions quite like those prior to the rise of complex animals and plants. None of the observed evidence suggested a single asteroid impact. Buick described the event in Australia as follows:

> Clearly, a single impact could not have been responsible. The most obvious interpretations are repeated environmental perturbations, such as methane hydrate melting pulses, repetitive overturn of a stratified ocean and/or persistent prodigious volcanic exhalations, or serial extra-terrestrial insults. Resolving which of these is the most viable explanation for the range of geological,

biological and biogeochemical features occurring over the extinction period is the aim of future research.

Serial extraterrestrial insults? Not even the Becker camp was arguing that more than a single rock from space was involved.

So where does this leave the impact hypothesis for the Permian? While popular science magazines such as *Discover* still promote the press-friendly impact hypothesis for the cause of the Permian extinction, among working scientists this is a rejected hypothesis. Once again, however, the all-too-common disconnect between what the majority of scientists believed and what a few media-savvy scientists believed led to very different points of view about the Permian extinction. The impact explanation continued to have support because of brilliant public relations work by the Becker team. Sooner or later, however, there would have to be new data if the conflict in opinion was to be resolved.

To settle the impact question, the Permian community took a page from the K-T days—use a neutral scientist as a referee to oversee the collecting of samples by proponents of both sides, and then have the referee randomize and distribute those samples to be examined, so that each side was testing some they had collected and some they had not, without knowing anything specific about their provenance or distance from the boundary between the two periods. Funded by NASA, a group that included Becker, Erwin, and impact specialist Frank Kyte of the University of California, Los Angeles (UCLA) as the neutral "referee" traveled in 2004 to the famous Chinese outcrop itself to see if the Becker results could be replicated. Small chips taken from the sampled rocks eventually were sent to various labs across the United States, including Becker's, but before the various labs could begin to analyze the geochemistry of these rocks, Becker and her crew quit the program, protesting that the wrong rocks in China had been sampled, even though she was there as they were sampled. The labs that completed their work found no evidence of buckyballs, no evidence of an impact.

THE FIRST EVIDENCE POINTING TO A PROCESS VERY AKIN TO THE REAL cause of the P-T mass extinction had by this time been known for nearly a decade. In 1996 a group led by Harvard paleobotanist Andrew Knoll published a startling new theory to account for the mass extinction, built on a realization that the end of the Permian period was much like the end of an earlier era, the Precambrian era, the time about 600 million years ago immediately preceding the advent of large animals and skeletons, and for much of his career, the focus of Knoll's research.

No one else was in a position to recognize a similarity between the two. Geologists typically concentrate their efforts on one narrow time, and this similarity came to light only when Knoll decided to jump to the Permian, taking his knowledge of the Precambrian time with him. Like the Permian, the Precambrian ended with large-scale swing in the ratios of carbon isotopes in the atmosphere and a mass extinction. Knoll and his colleagues argued that the cause of both was the same, and in their paper they proposed a novel explanation for how the changes transpired.

Knoll et al. suggested that the oceans of the late Precambrian era and the late Permian period were unlike those we have today—they were stratified, with water with more oxygen on top, and less below. Furthermore, these strange Permian oceans had large amounts of organic material locked in bottom sediments. Then, for reasons still unknown (but probably related to an increase in plate tectonic activity as well as a change in the continental positions), this pattern changed. The ocean somehow changed its state so that the deepwater, which had been safely locked away from the surface for so long, began to liberate its load of dissolved carbon, in the form of vast quantities of carbon and organic material, back into the shallow waters of the sea, and ultimately into the atmosphere, as large bubbles belched forth as though the oceans were a large soft drink. At the same time, one of the greatest episodes of volcanism known in Earth's history took place in Siberia, releasing more carbon dioxide directly into the atmosphere. The mechanism touted by the authors was akin to the horrific catastrophe that occurred in the 1980s at Lake Cameroon, in Niger, Africa. While thousands of humans and their livestock slept, the deep volcanic lake burped to the surface and into the air a gigantic bubble of carbon dioxide. This bubble spread out over the shoreline, killing most humans and animals there, before finally dispersing into higher altitudes, driven by winds. Was this the same mechanism that happened at the end of the Permian period, only writ much larger? Were all the oceans suddenly burping up bubbles of deadly carbon dioxide and other volcanic-like gases, such as methane?

Knoll and his colleagues proposed that the sudden increase in carbon dioxide, dissolved in the ocean, killed most marine species. Carbon dioxide in elevated concentrations is a known killer, and marine animals—especially those secreting calcareous shells—are particularly susceptible to carbon dioxide poisoning. The problem with this model, however, is that it cannot explain the coincident killing of land animals. Most terrestrial creatures are less sensitive to excess carbon dioxide.

There was a great deal of discussion both pro and con following this article. While Knoll and his colleague Richard Bambach began looking at various terrestrial organisms to see how susceptible they are to carbon dioxide poisoning, oceanographers, looking at the physics required to liberate large bubbles of carbon dioxide out of ocean water, could not get their computer models to corroborate that this event could take place. No oceanic carbon dioxide, no Permian event.

No one was saying that there was not copious carbon dioxide in the atmosphere—only that the carbon dioxide could not have been released quickly enough to kill anything.

Knoll's idea—that it was gas released from the deep ocean that was the cause of the Permian extinction—lay fallow for nearly a decade. But a variant of the same mechanism came forward in 2005, and it does present a plausible mechanism for compounds held in the sea to poison life on land. It was stimulated in no small way by the fossil record of land animals across the Permian–Triassic boundary. Much of these data were mine.

THE ROCKS AND FOSSILS THAT HAD FORMED THE BASIS, PRO AND CON, of the discussion of a Permian impact had all been deposited in the sea. But evidence for what was happening on land began emerging right about the time that the discovery of the alleged Bedout crater was announced in 2003. The evidence for methane release separately published by the Retallack team (for rocks in both Australia and Antarctica) and the Buick team (rocks in Australia) were both from strata that had originated in terrestrial settings, and new studies carried out in Greenland derived their evidence from fossil plants. Other evidence of fossil plant records appeared based on new studies in Greenland. And some of the most interesting new evidence of what happened to land animals came from studies that I had by then been conducting for nearly a decade in the Karoo Desert of South Africa, the most prolific fossil boneyard of late Permian and early Triassic age in the world.

For decades a succession of paleontologists has trekked into the vast wasteland of the Karoo to retrieve ancient bones. Early on, the fossils were collected with little regard for where they were found geographically, and with even less regard for their precise stratigraphic level. But since the 1980s, a new generation of paleontologists, led by

Roger Smith of the South African Museum and Bruce Rubidge of the University of the Witwatersrand, have brought rigor to the field. Their most recent work focuses on collecting just below and above the P-T boundary to test ideas about the severity and abruptness of the extinction among larger land vertebrates of the Permian period. Other, more novel approaches were brought to the Karoo by us American paleontologists who partnered with the South Africans to apply the kinds of isotopic studies that had been so successful at other geological boundaries, such as that at the end of the Paleocene epoch.

Two important findings emerged from this work: First, the isotope record showed not one perturbation but several. Second, while there was one interval when many species went extinct, it seemed to have lasted at least several thousand years, and there appeared to be smaller-scale extinctions both prior to and soon after this.

Thus, the land fossils seemed to show the same pattern as the marine fossils: a series of antibiotic events, the "insults" so colorfully described by Roger Buick. Maybe one of these was caused by impact, but by this time, teams from the University of the Witwatersrand, in combination with Christian Koeberl, another veteran of the K-T wars and an expert on impact evidence, could find no evidence of an impact at the levels in the Karoo where the highest rate of extinction seemed to occur. It's negative evidence to be sure—and no one to date has looked for buckyballs in this section—but nonetheless certainly not evidence of a single K-T–like asteroid strike.

THE WORK OF MANY INDEPENDENT TEAMS EVENTUALLY REVEALED A PIC-ture of the P-T boundary that seemed to show a succession of death intervals spanning a few million years before and after the event that marked the boundary. The picture was incomplete, however, because another line of evidence that had been collected from virtually every

P-T section had been largely overlooked. These were the carbon iso-tope studies.

At the K-T boundary, the isotopic evidence had demonstrated that a lush, plant-filled world went suddenly dead, remaining so for tens to hundreds of thousands of years. During that period, with so many plants and photosynthesizing marine plants having been killed off by the environmental aftereffects of the impact, the carbon cycle suddenly had a glut of carbon-12 that in happier days had been tied up in plants. But that was as far as this signal went: rapid catastrophe, followed by repair of the ecosystems and replacement of the killed-off species and individuals by the newly evolved and newly grown. Pretty quickly, things were back (at least from the carbon isotopes' point of view) to where they had been before the impact.

But a funny thing became apparent when similar kinds of stud-ies were conducted on late Permian and early Triassic rocks. At the P-T boundary there was indeed a perturbation indicating that plants rapidly died off. This was no shock—the fossil record had already convinced everyone that many plants had gone extinct. But that was not the end of it. Unlike the K-T, where the disruption of the normal isotope record was pretty rapidly healed following the one blow, the P-T record showed a succession of perturbations. If the world at the K-T were a boxer, it would be one caught unaware by a hard cross, knocked down but soon back on his feet—not overmatched, just sur-prised. The world at the P-T, however, was like a featherweight fight-ing Joe Frazier. Knocked down, it got up, just to be knocked down again. And again and again, as our foolish world kept evolving new plants (and animals, although it is the fossil record, not the isotopes, that tell you this), just to get its block knocked off again by whatever nasty boxer the P-T extinction mechanism (mechanisms?) really was. Now this was a surprise, and one that was hard to explain.

In a movie, at this point, some new brilliant scientists muttering *abracadabra*, among other mumbo jumbo, would appear on-screen, pull a slick new machine out of his or her sleeve (I would love to see Lara Croft take on the Permian), and solve the problem. "The Permian extinction is solved: It was caused by … caused by …"

But this is not the movies. By 2004 we were just beginning to find out what did not do it. To better understand the Permian extinction, still other mass extinctions had to be studied.

CHAPTER 4

The Misinterpreted Extinction

THE QUEEN CHARLOTTE ISLANDS, BRITISH COLUMBIA, JUNE 2001

B undled like mummies amid gear piled in the rear seats of the rattling, well-used helicopter—an aircraft piloted by a kid so young that it seemed questionable that he had a driver's license, let alone a pilot's license—four scruffy field geologists took off from the last outpost of civilization in British Columbia's Queen Charlotte Islands. The chopper was somewhat alarmingly overweighted, its four passengers obviously having not missed too many meals in recent decades, but even their combined weight was doubled by the mass of heavy gear lashed onto or squashed into every crevice in the JetRanger, an old Vietnam-era flyer. After the first lurch, rather than the normal slide into the air, followed by a slow circle of the logging camp helipad, the copter finally pointed westward toward high and craggy mountains that had to be crossed. The group was taken ever upward toward the snow-capped divide separating the well-logged eastern slope from the still pristine western parts of the huge island.

We had come to this isolated archipelago to visit a recently discovered outcrop of rocks spanning the Triassic–Jurassic time interval. Assuming the boundary between the two was not covered by seaweed or water, we were going to get a good look at the transition; the very expense and difficulty of the trip attested to how few of these T-J sections there were around the world. The goal on this trip was simple. The T-J boundary is marked by a mass extinction, and many paleontologists thought it might have been caused by a large-body impact on Earth, just like the K-T event. We on that trip had our doubts.

Whatever caused it, the extinction was enormous. More than 50 percent of marine animals may have died out, and as vertebrate paleontologists slowly gathered ever more fossil bones of this age, they too recognized an enormous extinction. The changeover on land was striking indeed. Prior to the extinction, the land world harbored a great bestiary of exotic reptiles and even a few mammals as well. There were many crocodile-like forms, and hulking herbivores called mammal-like reptiles, as well as the first primitive dinosaurs. But after the extinction, it was as if everything died out but the dinosaurs. How big was the Triassic extinction? Just possibly, in spite of all of the press that the Permian and K-T extinctions received, it was, in terms of the absolute numbers of species killed off, the biggest of all mass extinctions.

We were hoping to avoid our own extinction as the fragile helicopter struggled over the fanglike crest of the north-south mountain chain bisecting this large island, of a size to put Manhattan to shame, the helicopter's labors a stark reminder of how unnatural it is for a human to fly. A half hour more of flying finally brought us to the target beach, which had been discovered by Canadian geological survey teams. As isolated as almost any beach on the planet, this lonesome patch of black rock had long ago been named Kennecott Point, after the copper baron, some thought, but few had been the Canadians or Americans—other than the original Indian settlers of the Charlottes—

to visit this place. It would have remained but another obscure coast-line on an island filled with obscured coastlines, if not for the rise in interest in mass extinctions.

After doing one full circle over the dark beach and darker water, the helicopter dropped with stomach-turning speed toward one of the few places on the rocky beach flat enough for us to land—in a high, buffeting wind, the pilot had decided to rid himself and his craft of this heavy human and equipment payload as quickly as possible and make haste back to the safer side of the island. Over the intercom he informed us that he would keep the copter going during unloading and that we should watch our heads, as in keep them attached to our bodies. A thumping, back-wrenching landing, and the four of us tossed our gear onto the wet beach in the rotor hurricane before ignobly scrambling out amid the scattered gear. Stumbling higher up the beach, we were sandblasted by the rotor wash as the copter jumped skyward, freed of the heavy cargo of its outward voyage, its staccato noise echoing up against the nearly vertical mountains that pinned this beach against the sea. Soon enough that last reminder of human industry was lost in the keening noise of the ocean's wind against the old-growth cedars as the raging Pacific storm—quite normal for a place that gets 200 inches of rain each year—unceremoniously greeted our group.

We humped the heavy gear through the monsoon into the nearby forest: food, water, sample bags, drills, and hammers—and even a small portable drill rig. Tents went up, but not fast enough under the dripping trees, sleeping bags getting the first damp that would never leave them for the week we were there. We were far enough north that daylight was an 18-hour proposition, on a beach where the next stop west was Japan, and the only inhabitants were deer, dead Haida Indians in their graves among the rotting village back in the trees, and some really large and crafty black bears that surely lived just behind every bush that we geologists might commandeer for a toilet. But the real enemy at this site was not bears. It was the rain, enemy to the portable drill, enemy to the so-called write-in-the-rain field notebooks, enemy to the Gore-Tex that was so quickly overwhelmed by the rain, enemy to the geologists' mood and cooking and sleep. It was a rain that brought to life the smells of the hundred men who had lived in the standard-issue Geological Survey of Canada (GSC) mummy bags we each carried—our trip was sponsored in part by the survey, and as an old organization, it had traditions to be honored and rules to be followed.

Awaking at 4 AM to GSC mush (oatmeal and chopped apples, with a side of instant coffee), we scrambled into wet clothes and trudged onto the outcrop at 5 AM. Long days ended up seeming longer, made so in no small way by the lack of immediate gratification, for the target strata—dark as death, deposited on a deep late-Triassic ocean bottom that seemed nearly devoid of life and that certainly was nearly devoid of fossils—offered no morale-lifting moments of fossil discovery, where not only is something beautiful uncovered but knowledge is immediately gained as well, pointing to new understanding. No, this trip was about measuring great piles of almost fossil-bare strata, smashed nearly vertically, on a beach controlled by tides. The tides controlled our work, too, and high tide was time for naps or for looking for glass floats from 1955-era Japanese fishing fleets.

All of this took time. Outfitted in yellow or orange rain slickers and great green gum boots, from a distance we looked like children on a school playground on a rainy day, but we were four dripping souls trying to write, sample, or find fossils in unceasing rain that made black rocks even darker.

We had made a previous trip here in 1999 and had put small concrete monuments into the rocks. The markers had been set up at 10-meter intervals so that any rock samples taken or fossils subsequently found could be allocated to a specific stratal horizon. To our surprise, we found that the majority were gone, after two winters, testament to the ferociousness of the North Pacific storms that smashed this beach 9 months out of 12. So we had to again undertake the onerous task of measuring and marking the sedimentary beds, so that every sample would have a known position in meters above or below the mass-extinction bed. Except that there really was no mass-extinction bed. We had to guess where it had happened in a 10-meter thickness, bracketed by the last Triassic fossil and the first Jurassic fossil.

Despite the storms, at the end of the week there were stacks of fossils and sample bags crammed into giant metal cans, all lined up for helicopter extraction, and the first distant noise of the arriving helicopter was met with relief. Food had run out the day before as the perpetual storm worsened to the point that even the Vancouver Island Helicopters' *Apocalypse Now* pilots were grounded. The helicopter dropped down, and now it was a mad scramble to pack it for the return to civilization.

AT THE TIME OF OUR FIRST TRIP TO THE QUEEN CHARLOTTE ISLANDS IN the last summer of the 1900s, there had already been many futile attempts, in many parts of the world, to see if the Triassic mass extinction was accompanied by the same isotopic perturbation that had been observed at the end of the Cretaceous, Paleocene, and Permian time intervals. But the question was not only whether there was a noticeable perturbation in the ancient carbon cycle at the end of the Triassic but also what kind. Whether it was marked by one large change in the carbon isotope ratio—as was found in rocks right after the K-T catastrophe, when the surface plankton was destroyed but the deeper organisms remained relatively unscathed—or by a series of perturbations toward both "lighter" as well as "heavier" carbon isotope values that by 1999 were known from the Permian and Paleocene events was a question of paramount concern. Each attempt so far had yielded gibberish for results, the rocks having been chemically transformed from their original conditions.

Not expecting much, after the summer of 1999 our small group had prepared our samples from the dark beach and fed them into the maw of a mass spectroscope housed in Seattle. We had already expected that there would be a perturbation in the carbon isotope values—the fact that there was an extinction at all suggested as much. And sure enough, our carbon isotope results from that summer of 1999 revealed a beautiful carbon isotope record of the extinction. (One of the reviewers for our paper, ultimately published in *Science*, was an elderly British Triassic specialist who sniffed, "Ward et al. struck lucky." Well, if it was luck, we took it gladly.) But one single isotopic perturbation was a K-T signal. Yet because our sampling in 1999 had not gone very high above the extinction level, we had no idea if the youngest Jurassic rocks would show the multiple carbon isotope perturbations that by the first years of the twenty-first century were known to characterize the Permian extinction.

Thus this second trip in 2001—to sample higher in the section, concentrating on strata in the Jurassic. If this section showed a Permian pattern, it would be a strong indication that the Triassic extinction was allied in cause to the Permian extinction—and that neither was caused by impact. If the single isotope perturbation was all that was there, however, it would be a vote for impact.

Little by little, results came back. Months after our collection, on a day pouring with rain, Seattle rain this time, enough numbers had come from the mass spectrometer to allow a high-resolution look at the relative amounts of carbon-12 to carbon-13 from samples taken at one-meter intervals from the highest Triassic, across the boundary, and then more than a hundred meters up into the Jurassic. There was the already discovered perturbation right at the boundary, all right, but then the results showed another, even larger one going in the opposite direction isotopically, followed by another negative excursion. Extinction, rebound, second extinction. This was a Permian-like signal, all right.

IN EARLY 2002, JUST AS OUR NEW RESULTS WERE CONVINCING US THAT the Triassic mass extinction was more like the Permian than the K-T mass extinction, a sensational paper arrived that seemingly negated our conclusion. With great press fanfare, it announced that evidence of impact had been found at a new T-J boundary site, in this case located near Newark, New Jersey. The rocks being sampled were not of marine origin, like the first K-T, P-T, Paleocene, and even our Kennecott Point rocks had been, but instead came from sediments that had formed in a great rift valley that came into existence when Europe split away from North America in latest Triassic time, an event that brought the Atlantic Ocean into existence. Before it could be flooded by the ocean, however, a series of valleys as long as the east coast of North America filled with shallow lakes, probably looking like the East African Rift Valley lakes, and, like those African lakes, the lands around the ancient Triassic ones must have been home to unbelievable numbers of large land animals. Unlike modern Africa, however, so filled with mammals (and plenty of crocs too), the ancient Newark Valley lakes must have been home to large numbers of dinosaurs, judging by the spectacular assemblages of footprints that came to be preserved in rocks that would someday make the brownstones of the many cities in the New York City region. For more than two centuries it has been known that the many rivers and creeks in the Connecticut River Valley and Newark Basins are home to the most diverse assemblage of late Triassic and early Jurassic dinosaur footprints on the planet. The new paper combined mention of dinosaur footprints with an announcement of hallmarks of impact—iridium and the rest—into a spectacular splash in both scientific and the popular presses.

It was, of course, this association of dinosaurs and mass death that whetted the journalists' appetite for extensive press coverage in the first place. So once again in mass-extinction work, new results seemed to swing the sum of evidence back toward impact as the dominant producer of mass extinction, just as it had 20 years previously for the K-T mass extinction. The lead author of the paper, published in the flagship journal *Science*, was Paul Olsen of Columbia University, a paleontologist who had spent an entire career working on the T-J boundary.

Olsen and his colleagues reported an iridium anomaly from continental T-J boundary beds in several localities across New Jersey. It was, of course, just this kind of anomaly that had first alerted the Alvarez team two decades earlier to the possibility of impact at the end of the Cretaceous period; iridium had become the gold standard of impact evidence. But beyond this, Olsen's study wildly diverged from the template of the 1980 paper by Alvarez et al. Where the Alvarez group followed the physical and geochemical evidence from its Italian boundary section with data confirming mass extinction of small ocean life at the same time as the impact, the Olsen paper for the Triassic event followed the physical and geochemical evidence with just the opposite result: Olsen's team found that rather than eliminating most life in its section, instead, the impact seemed to have acted like a biotic fertilizer, leading to both more and bigger life!

Granted, the kinds of sedimentary rocks studied as well as the fossils enclosed could not have been more different. Where the Alvarez group reported on the fossil record of tiny protozoan shells deposited on a deep ocean bottom, the Olsen group sampled strata deposited on land (or, more correctly, in streams and shallow lakes on land), and the fossils it studied were footprints, not the remains of body parts. But in spite of these rather startling differences, Olsen concluded that a great asteroid had hit Earth (this time about 200 million years ago, the age of the T-J boundary) and that as with the K-T event, the dinosaurs were affected. But from that point onward, the conclusions could not have been more different! Alvarez et al. argued that an impact killed off the dinosaurs. Olsen et al. seemed to suggest that their newly discovered impact acted like dinosaur fertilizer: Right after it hit, they argued, there were more and bigger dinosaurs than before the "impact." Reading this for the first time, many of us could only mutter the equivalent of *wow*. I remember thinking that, to paraphrase the Bard, something was rotten in Denmark, or in this case, New Jersey. Paul Olsen was no Luann Becker, shrouded in secrecy. Au contraire—to try to better establish his new discovery, he brought in all who cared to look to his urban outcrops. Plenty of the many specialists working on mass extinctions at the time made the trip.

NEWARK VALLEY REGION, NEW JERSEY, JUNE 2002

Olsen organized and led a field trip for two dozen geologists. His intentions were noble: to show all and sundry where he had conducted his sampling and even to invite others to take their own samples. The trip began on an urban note. An endless rush of New Jersey residents drove the freeway nearby, creating enough noise to require Olsen to shout to the geologists gathered in a parking lot behind a decaying strip mall, real Tony Soprano territory. The assembled group, all dressed in geology field gear, presented a somewhat incongruous sight to passersby as it followed Olsen to the low outcrop of sedimentary rocks behind the stores. It was not the usual promenade from field vehicle to outcrop. The ground was strewn with old garbage sitting on much older rocks. In one corner the pungent smell of urine brought attention to the broken plastic and glass of syringes and crack pipes.

To those from the West, where looking at rocks was synonymous with being in outdoor wilderness, this particular stop on a two-day bus ride across Triassic-age rocks making up much of New Jersey was a surreally bad geology dream but one that future geologists had better get used to as cities and suburbs spread ever larger across the land, running over rocky outcrops of interest to geologists in the process. This was one of the new orders, it seemed.

We geologists moved closer to the red sedimentary rocks making up the back of the parking lot, irrationally trying not to touch anything. Olsen directed us to the strata, and we looked with more interest (everyone loves to find a fossil) at the fallen talus at the base of this outcrop in light of the treasures that he assured everyone were there: in this outcrop, fossil fish that had once lived in the Newark Valley's rivers and lakes, and in the next, dinosaur footprints, something that few of us had ever seen on an outcrop.

The ancient lakes yielded plants and fish, while the river deposits held other treasures; some bones were there, but it was these dinosaur footprints that primarily interested all on the field trip. In many areas the red mud lining shallow ponds and swamps near the lakes and rivers were the feeding and drinking places of a diverse suite of large animal life, most being early kinds of dinosaurs. Most of these were bipeds, miniature versions of the much later *Tyrannosaurus rex* and its relatives. The day wore on, and all of us began to ripen in the humid bus under a scorching East Coast sun. Outcrop after outcrop began to get exhausting, and each person piling back into the bus required a partner to remove the numerous ticks that constantly fell on us as we passed through leg-high bushes or beneath trees where these monsters must have been swarming, waiting for mammal blood and victims to unknowingly give Lyme disease to.

The geology was spectacular, and to many of us, novel. After seeing many examples of late Triassic sediment, I was surprised to see us head into the Palisades region lining the Hudson River. Here the massive cliffs were not sedimentary rocks at all but giant heaps of frozen lava, and the point brought home as nothing else could that while the Triassic was coming to an end, Earth itself was engulfed in a paroxysm of volcanism. While Olsen wondered at this coincidence, that an asteroid had hit just at the peak of volcanism spanning the Central Atlantic Magmatic Province, a flood basalt deposit stretching from Brazil north to the Bay of Fundy, I wondered at the magnitude of that volcanism and the volume of carbon dioxide it would have produced.

We had all long known that the Permian ended at the peak of volcanism forming the Siberian Traps, but none of us had traveled to that faraway place to see piles of Permian lava. Our understanding of volcanism and mass extinction at the end of the Permian was an intellectual one. Here, at the end of the Triassic, was the in-your-face fact of major volcanism coincident with a great mass extinction not just in the sea, among the fossils typified by the Nevada jaunt that began this book, but among land animals as well, whose footprints had made the *Science* paper by the Olsen group all the more remarkable.

It was late afternoon when the bus arrived at the last outcrop of the field trip, and Olsen had saved the best for last. Again, compared with more Western standards (or even European standards, for that matter), the outcrop—a road cut leading to a new housing tract—was pitiful. Lush vegetation of New Jersey largely covered the modest, ten-foot-high wall of rock that held Olsen's prize locality. Running down the middle of the outcrop was a black band, a thin coal seam. Right beneath this, Olsen's samples had yielded iridium, and, unlike what had happened to the Becker work, various labs confirmed his findings. But there was still the nagging worry. One would think that the footprints after the Triassic mass extinction would be fewer in number with fewer animals around to make them, fewer in kind as species died out, and smaller in size, since one lesson learned from the asteroid-caused Cretaceous extinction is that it was disproportionately lethal to larger animals.

Just as with the Becker work of a year prior to publication of Olsen's work, those of us working on mass extinctions, impacts, or both scrutinized the Olsen paper in painstaking detail. Within a day, the consensus of various scientists with experience from the study of the K-T event was that the iridium evidence at hand, let alone the strange footprint evidence, did *not* support a K-T–like impact. This was certainly the opinion of two experts on interpreting impact deposits, Frank Kyte of UCLA (whom we met in Chapter 3, "The Mother of All Extinctions," as the "referee" of the Permian blind sampling program) and David Kring of the University of Arizona. While both were of the opinion that the iridium finding was certainly indicative of an impact about that time, both also pointed out that the amount of iridium reported from the various sites of the Olsen group was at least of an order of magnitude smaller than that found at virtually every K-T boundary site. Something fell to Earth, all right, but it was small, probably too small to cause the amount of extinction at the end of the Triassic.

This was not welcome news to Olsen. But any number of reasons could be found for the low iridium values in addition to the most logical, that the asteroid was much smaller than the Chicxulub rock at the end of the Cretaceous period, so Olsen pressed on with his claim. And he had a really good candidate for causing his hypothetical impact event sitting in serene splendor in buggy Quebec. There, long known, was one of the biggest craters visible on the planet—the Manicouagan crater—with a diameter of about 100 kilometers (in comparison, the Chicxulub crater is 180 to 200 kilometers in diameter), plenty big enough to cause the end-Triassic mass extinction. It had long been thought to be of the right age too—somewhere near 210 million years in age, which was about the age of the T-J boundary. But then this exciting possibility began to evaporate.

In 2002, Jozsef Palfy, a Hungarian student working on his Ph.D. at the University of British Columbia in Vancouver, collected ash samples from a T-J boundary in the Queen Charlotte Islands. Although this was not from the locations where my team had worked, it was from rocks with similar fossils and gave us a really good date for the formation of the boundary. But the result was a bit of a shock: The radioactive decay measures indicated that the Triassic period came to an end about 199 million years ago, a date later revised in 2005 to 201 million years ago. And not only did the end of the Triassic get younger, but the Manicouagan crater got older. Better dating placed its formation at 214 million years ago.

With this date in hand, many turned to the literature, or their "own" stratigraphic sections, to see if the interval deposited about 214 million years ago could be found, and if so, if anything went extinct. Very few strata from any point in Earth's history have dateable ash layers, and such was the case for most Triassic strata. No one could find ashes in fossiliferous sedimentary rocks of exactly this Triassic age anywhere. But we could make estimates on the basis of the types of fossils present, and the result was believable: There are no mass extinctions in the record during a long interval of the late Triassic. This finding had ramifications far beyond simply the study of the Triassic. Here was a large crater, which David Raup had earlier estimated should have been caused by an asteroid large enough to kill off between a quarter and a third of all species on Earth, and we found nothing! Nothing happened! The lethality of asteroid impacts might have been overestimated.

There remained one more bit of work to do, but it was the hardest. What was the record of extinction at the end of the Triassic? Was it sudden, like the K-T, or spread out, like the P-T?

We began this chapter with the possibility that the Triassic extinction was the most catastrophic of all, and right now that cannot be dismissed. But whether first, second, or third, it remains a catastrophe far beyond biblical proportions, to steal a phrase. But how long did it take for this catastrophe to unfold? Only the fossil record, combined with new and accurate ways to date sedimentary rocks, could answer this.

QUEEN CHARLOTTE ISLANDS, 2004

Once more jumping out of the helicopter, the third time here. Yet another kid at the controls, and as I scrambled out, I saw him gesturing frantically at me. Later, back at the base, I heard that he grumbled that the "old gentleman" had almost had his head cut off from the rotors while unloading, but in the game of helicopter rotor decapitation, a miss is as good as a mile. (*Old gentleman?* Just because I had grayed and moved with a very distinct limp? Who was he calling old?)

This third trip had a specific goal—to get a better view of the fossil record here and to try to see how long or short the extinction had been. Only then could we know if there had been one extinction or several and whether it (or they) had been gradual or sudden. To find out, we needed to get to a small island offshore of Kennecott Point, where rocks slightly older than those on the point itself could be found. The oldest rocks at Kennecott were upper Norian stage in age, the second to last stage of the Triassic. We needed to get lower in time than that, but to do so we needed a Zodiac boat with a small outboard motor to get there, and so to carry that weight the copter we had ridden in on was far larger than any we had ever used—a Huey this time. I had been humming the Wagnerian song from the movie *Apocalypse Now* as we came in the first morning; there was the smell not of napalm but of déjà vu.

My crew was a good one: Jim Haggart, my first Ph.D., dating back to my days at the University of California, Davis, and now the specialist for the Cretaceous period for the Geological Survey of Canada; Chris McRoberts, world expert on Triassic bivalves and a man who had seen more T-J boundary sites on Earth than anyone else, all except this one; Geoff Garrison, my postdoc student at the time and the man who ran all those crucial isotope analyses; Ken Williford, my grad student, who was studying the Triassic extinction of New Zealand, Nevada, and British Columbia; and Isaac Hilburn, a grad student from Cal Tech sent to us by my long-term research partner, the great Joe Kirschvink, my coauthor on Permian and now Triassic work.

The gods smiled on us. We camped, as usual, under the tall trees near Kennecott. But this was July, and maybe because of global warming, maybe just by luck, the sun warmed our week.

Everyone had his duties, and the team was like a well-oiled machine. My job was to be the overall scientific pooh-bah, but because Haggart, one of the great field geologists of all time and a man whose middle name was *Organized*, had set up the trip, as he had all the previous trips, and because my crew, all good scientists, knew exactly what they wanted to do, I needed to do little leading. I was happy to be taken to my field region, in the Norian rocks on a small island offshore of Kennecott (the Norian stage is the unit of time prior to the last stage of the Triassic, the Rhaetian), to look at the extinction of ammonites in this section.

My boys were up to the task. By boat Haggart would, in two trips, whisk through the bobbing kelp beds, diving seals, wheeling seabirds, and stormy chop of the cold North Pacific Ocean to Frederic Island, our tiny target-islet a quarter mile west of Kennecott and the last stop of land between North America and Japan. Garrison and Williford took isotope samples or helped Hilburn in the time-consuming job of extracting oriented cores with a modified, diamond-tip coring chain saw, for magnetic stratigraphy studies; McRoberts collected clams.

Having this much labor along made life easy. My legs no longer worked very well by 2004 because of the abuse they'd taken from the many rocks I had fallen on since 1981, the many barbed wire fences I'd jumped over, the miles I'd trudged in the dust, the many rock walls I'd scaled and fallen off of—but mostly, in all probability, from a diving accident in New Caledonia in 1984, when my field assistant at the time died after passing out at 200 feet deep, requiring me to haul him up without decompression steps in a vain attempt to get him back into our natural air environment, an emergency ascent giving me the bends in both knees and my left hip, which would ultimately force me to replace it with titanium and ceramic in September 2006. I had necrosis in the bony parts of my leg joints, and no amount of ibuprofen or even Vicodin was going to change that. I had to drag myself over a few outcrops, sometimes with help like a sack of wood, to the good-natured insults and amusement of my crew, who endlessly repeated the (to them) hilarious observation that on this trip I was not "carrying my own weight," an observation that even over a week apparently never got old. In fact the fastest that I moved on the whole trip was the last afternoon, when Hilburn, the grad student from Cal Tech, thought it would be funny to dispose of the excess white (stove) gas by throwing the whole sealed gallon can into a giant fire where we were burning our refuse, rather than draining it as instructed. While this should have been a Darwin Award moment, the ensuing explosion and shrapnel destroyed only nearby trees and probably kept the bears at bay for the rest of the trip. (The Darwin Awards salute the improvement of the human genome by honoring those who remove themselves from it.)

That notwithstanding, I had the best job. Not the boring drilling of rocks for magnetostratigraphy. Not the equally boring sampling of small bits of rock for eventual isotope studies. Not even the patient work of measuring the stratigraphic sections, all missions on this trip.

My job was right in line with the reason I went into paleontology in the first place: to find fossils and to accomplish the aforesaid reason for this trip—to find out if, at least in this one locality, the Triassic extinction had been fast or slow, single or multiple.

For several perfect, blue-sky days I got to slowly work my way up, layer by layer, ever higher in the Triassic strata. I would find some large bedding planes, the large expanses of the very dark strata where angled afternoon sunlight could reveal the ribs, tubercles, and shell shapes of numerous small ammonites smeared on the stratal sheets. There were so many! They were mostly small, but quickly I began to see the variety of them, so many species here in addition to individuals. These rocks were deposited about three million years before the Triassic "extinction," and as my notebooks and sample bags filled with the written or stony evidence of a seemingly ammonite-packed world well before the T-J mass extinction, I could see how rich with life this sea bottom had been well before the extinction.

Slowly I worked up through time. The bedding planes began to disappear, increasingly covered by gravel, logs, the wrack of beaches exposed to the great gales of autumn, winter, spring, and, I guessed, summer on occasion as well, even if not this summer. So it was mining on a small scale, hammer and chisel into the black strata. My pits began to cover the beach, and the numbers of ammonites revealed a true gradual extinction, a pattern not at all like that unearthed on the Bay of Biscay beaches.

Irony is a funny thing. From 1981 to 1992, I had similarly mined the ammonites of the uppermost Cretaceous period. Eventually, those efforts showed that K-T was not a gradual but a sudden extinction. Here, on this beach in British Columbia, just the opposite pattern showed. Meter by meter higher in time, the number of kinds, as well as the sheer number, of ammonites dropped. By the top of the beach there was but a handful of species. And the ammonites were not the only animals to disappear. Near the top of the Frederic Island section there were rocks absolutely packed with fossil clams. The same rocks were found on nearby Kennecott Point, as well as many other places around the globe. The clams were named *Monotis*, and they, and the few remaining ammonites, could be seen to gradually disappear in numbers to total extinction—but their extinction occurred a hundred stratal meters below the end of the Triassic period.

If this were the whole story, it could be argued that perhaps it was but collecting failure and the vagaries of sedimentation and the fossil record, and that, as was the case with the Cretaceous story, those complications made a truly sudden extinction look gradual. But there was an enormous difference. The disappearance of the many *Monotis* clams, and the few ammonites remaining with them of the hordes that lived only tens of meters before, or some million years prior, did not mark the end of the Triassic period. It only marked the end of Norian stage.

A pattern emerged. In the upper part of the Norian stage, perhaps 212 million years ago, the sea was seemingly packed with creatures that would enter the fossil record. The ammonites were chief among these. But as the Norian waned, two things happened. First, the ammonites dwindled down to only a few kinds. Second, right at the top of the Norian, a new kind of clam appeared in such profusion that it packed the rocks with its fossils. This was *Monotis*, and its appearance here might indicate that things were getting bad in this ocean, for *Monotis* seems to have been adapted for low-oxygen bottoms. And then even these clams disappeared, and we could track that disappearance. Over several meters of strata, they became smaller and fewer in number, and then they were entirely gone. Welcome to the last stage of the Triassic, the Rhaetian.

From this point, the base of the Rhaetian stage, to the T-J boundary, there were about 100 meters of stacked beds. But they were curious beds, very different in appearance from those of the Norian stage. They were striped between gray and pitch black. They had trace fossils, but of a kind that signaled low-oxygen bottoms. There were a very few ammonites, and these too were curious. While all of the Norian ammonites were tightly coiled, these last Triassic ammonites had completely uncoiled, to form long, straight tubes that we interpret as adaptations to a floating existence on the surface of the sea.

So this long stretch of strata looked like a dead world. But how long did it last? The guess of the geological community was that the Rhaetian stage lasted between 3 million and 6 million years. But here is where the new kind of rock dating comes in. We found and collected an ash layer from the last beds with *Monotis*. Long ago, while the *Monotis* were dying out, a nearby volcano blew its top and sprinkled ash over ocean and land. It sank to the bottom of the sea, to form a half-inch layer of volcanically derived particles. It is these that can be dated, and in 2006 that data on the date came back. This ash had been deposited about 211 million years ago, give or take 1 million years! Because the T-J boundary had already been dated at 199 million to 200 million years in age, that meant that the Rhaetian stage had actually lasted as long as 12 million years! I was flabbergasted. For 11 million years, it seemed, the world was already dead but for its oceanic plankton. The ammonites and clams were largely gone from the Kennecott Point strata. When the T-J extinction finally came, its main victims were the one or two species of ammonites still present, as well as most of the plankton.

Thus the nature of the Triassic mass extinction came into view. For millions of years near the end of the Triassic period, the fossil and rock evidence from this and other places indicated that at least the marine portion of the Earth was a place of very little life. The date of the final nadir of this dying world in the oceans occurred some 211 million years ago, as noted above. However, the lack of ash beds precludes knowing whether land life died out like marine life, slowly, in pulses. The end of the Norian stage marked death in the seas and perhaps on land as well. The Rhaetian stage was a time of little life, and what little there was, was finally all but snuffed out at the end of the Triassic period itself. The Triassic extinction is at least two extinctions, one at the end of the Norian stage, when the bivalves and most ammonites disappeared, and one at the end of the Rhaetian stage. But that was not the end of the matter. The isotope perturbations that we and other labs had by then found from many places around the globe in the first stage of the Jurassic period, the three-million-year-long Hettangian stage, showed that small pulses of extinction continued well into the Jurassic.

THE GREAT MASS EXTINCTION THAT ENDED THE TRIASSIC PERIOD THUS turned out to look nothing like the K-T. Given that the Paleocene, then the Cenomanian–Turonian, then Permian, and now the Triassic extinctions were known to be fundamentally different from the K-T extinction, it is no wonder that some geologists began to doubt the paradigm that all or most extinctions were caused by impact. The clues to the true cause of these mass extinctions would be scattered not only in the ancient fossil record but in the modern world as well—in the ocean and its current, in our volcanoes, and even in noxious lakes and ponds on sun-drenched islands in Micronesia.

CHAPTER 5

A New Paradigm for Mass Extinction

THE SHORT DROP-OFF, PALAU, JULY 1983

A pair of divers slowly rose up the side of the blue wall, into the zone of living coral and red colors now, into schools of fish. The two men—I was one of them—floated upward no faster than their slowest bubbles, short inhales and long exhales from the rubber mouthpieces clutched by experienced mouths. Each of us had one arm stretched upward, making us look like two eager schoolboys trying to provide a well-known answer to some question, but in this case we each also had a hand gripped on the flexible hose extending from our buoyancy compensators, for our ascent from well below 130 feet demanded a bleeding of the buoyancy compensators, the gas in our lungs, gas in our tanks, gas in our blood, and, most dangerous, gas in our flotation devices expanded from the lowering pressure dictated by Boyle's law.

At 40 feet we came to the top of the sheer coral wall, confronted by the enormous expanse of the reef front and reef top itself, a warm, sunny, multicolored universe of unbelievable diversity, of unbelievable abundance, the marine equivalent of a rain forest, but a delicate rain forest, one balanced on narrow ranges of heat and oxygen. Strong leg kicks powered us over the edge of the wall into this vast summery habitat, into ever shallower crystalline water, the anchor line of our boat now brightly visible, a yellow line pointing up toward the world of air we would soon return to. Soon, but not yet. Surfacing now would be tempting fate, physics, and physiology, which demanded that two decompression stops be made, for a couple of minutes at 20 feet, and for a good 10 minutes at 10 feet.

Our dive had been so deep to supposedly find deeper water fish for the American continental aquarium that my diving partner, Michael Weekley, worked for; the reality was that this was a pleasure dive, at a beautiful reef wall somewhat egregiously misnamed the Short Drop-Off. There was nothing short about this reef's drop, with depths exceeding 2,000 feet only a few hundred yards from the top reef's breaker zone. This wall was also the closest to our month-long base and sometime home: the Palau Mariculture Demonstration Center in Koror, the jewel of Micronesia and site of some of the most luxuriant and pristine coral reefs in the entire Indo-Pacific region, the vast swath of the tropical Pacific Ocean that is the diversity center of coral, mollusks, and tropical fish on our planet.

Our trip as a whole was about more than collecting fish for aquariums. We had come to Palau to further unravel the mystery of cephalopod biology and paleobiology, including that most interesting question concerning the relative fates of the two great stocks of externally shelled cephalopods, the ammonites and nautiluses. The ammonites had survived many mass extinctions, even if wounded. Through the end of the Permian period, end of the Triassic period, and through the several other extinctions during the Mesozoic era, they survived to flourish in ever-greater numbers following the events. Not so during the Cretaceous period. The ammonites died, quickly, while the nautiloids had survived, and lived still, in large numbers in that water. But the most curious aspect of it all was that the nautiloids seemed to show a very different pattern of survival compared with the ammonites at each of the mass extinctions—at the end of the Permian and Triassic, more nautiloids died out than did ammonites, but at the end of the Cretaceous it was the ammonites that went bust—and we scientists thought we might be discovering why. Somehow the surface waters were as lethal at the end of the Cretaceous period as the deeper waters had been during the other mass extinctions.

Lethal? Well, not exactly these waters, not today, I mused, holding on to the rope with clenched knees, keeping my depth at the 20-foot mark on my depth gauge, watching the seconds tick by on my watch, hoping all the nitrogen molecules that had been forced to dissolve into my blood by the deep dive were now making their way out of solution and back into gas in my lungs, and not into my bloodstream, liver, brain, or spinal column. That gas could cause death, hideous death, simply by changing depth was a curious thought, and although there is nothing more boring than hanging on a line, there was no going up to that nearby boat—we were unable to surface even if the Loch Ness Monster were hungrily circling (or, more to the point, if a large tiger shark were circling, which had happened to me while decompressing after a deep night dive photographing wild nautiluses in New Caledonia the year before).

Two minutes were up, and we moved even shallower, into the blood-warm 10-foot-deep water, the boat bottom quite close now, things looking closer in water than they were. Ammonites and nautiloids—one stock lived and one died—and I was having vague ideas about why this curious pattern had occurred. At the time the Alvarez impact hypothesis was the hottest science on the planet, and a major aspect of the new research dealt with finding out which organisms survived and which perished, and why. My trip to Zumaya the year before had put me into the middle of this controversy. Some of the paleontological silverbacks of the time espoused simple chance—it was not bad genes, simply bad luck. There had to be a reason that the nautiloids made it, to survive right up to now, as evidenced by the huge populations of them here in Palau and elsewhere in the South Pacific, and I suspected that I knew.

Over the course of the past four weeks we had pulled off a wonderful scientific feat, or so we congratulated ourselves in the local bar afterward, drinking the only beer then available, Old Milwaukee. Four of us had come to Palau armed with miniature transmitters that could relay information on the motion and depth of any animal tagged with them—if one could stay in range of the battery-powered transmitter, that is, and that range was less than half a mile in the best of circumstances. We had caught a large cage full of nautiluses the first week of our stay, and we had tagged four of them by permanently attaching the small transmitters to their shells.

After that came a nightmare time—tracking meant sitting over the deep-living nautiluses in a small boat, and the only boat that was safe enough to move around the treacherous reefs at night was an open Boston Whaler with a big outboard engine. Even marking the dangerous reefs each sunset with glowing Cyalume-filled markers only partially alleviated the chance of running onto the wave-swept reef tops in the dark of the tropical nights.

The tags on each of the four nautiluses gave out continuous information, with a strain gauge causing their sonic beeps to increase in duration with ever-greater depth. But there was no automated recording of this data; instead, every 15 to 30 minutes, a hydrophone was dangled over the side of the boat, a measurement of the depth was taken, and a precise location was noted on the nautical chart of the region.

We took turns sleeping in the bottom of the boat, huddled together under a tarp during the frequent rain squalls coming out of the endless Pacific Ocean, took turns running the even smaller lifeboat to shore to pick up more sandwiches, took turns climbing into the sea to urinate and defecate, took turns slathering endless quantities of sunscreen onto cracking tanned skin, took turns telling our life stories or dissecting our relationships with girlfriends and wives yet again.

Seven days and seven nights, we followed one particular tagged nautilus, getting seven days and seven nights of data for that one and at least three days' and nights' data for the other three. From these four nautiluses we found that each day these cephalopods dived away from the sun's light and spent the day in slow motion in the darker depths of up to 1,500 feet. But each sunset, as night fell with tropical swiftness, they swam inward, up the contour toward the reef, never into water shallow enough for a human but far shallower than their daytime habitats, moving and feeding at 400 to 500 feet, always on the bottom. It was clear that these animals were part of the deepwater fauna, not members of the sunlit world.

Depth. Was that the reason for the survivability of the nautilus during the Cretaceous crisis? It was already known that the nautiluses lay large but few eggs, and that these eggs take a year to hatch. Through the use of oxygen isotopes it had been found that the hatching depths were more than 700 feet. Ammonites, on the other hand, seemed to have been dwellers of far shallower water. If the ancient nautiloids were similar in habit to their still-living counterparts, the puzzle at least had all of its pieces, if not in their right places. At the end of the Permian and Triassic periods, the deepwater animals fared worse, while the shallower forms did better. At the end of the Cretaceous period, however, it seemed that just the reverse held. The shallow-water fauna was almost exterminated, plankton as well as animals, but the deepwater forms—the diatoms and the nautiloids—came through unscathed. Those researchers studying the effects of asteroid impact, work catalyzed by the Alvarez hypothesis, came to the conclusion that the surface of the sea down to 100 feet would have been lethal to most of its inhabitants, owing to a combination of high acidity and toxins falling from the sky after the titanic impact. The ammonites lived up there, bred there, and at the end of the Cretaceous, died up there. Yet in the other mass extinctions, it was as if just the opposite held true: In those, such as the Permian and Triassic events, it was the deep that was more lethal than the shallows.

Five minutes left on the line at 10 feet, which meant only five minutes more to visit this unbelievably beautiful place, for this was the last dive, the last day, and after it was over, it would be time to break camp and load up for the long trip home scheduled for the next day. I reflected on the day before, another kind of dive, one in a place very different from this one. Our team had visited a large, baking, and stinking freshwater lake in the interior of the island, one famous for the untold numbers of jellyfish that floated in the crystal-clear surface waters of the lake. But we found that the water filled with jellyfish was but a thin stratum atop a very different water mass. Below was a place with no animals, for it had no oxygen. What it did have was a deep purple color, and rising from this deeper layer of far more primitive life were small bubbles of a toxic gas: hydrogen sulfide. The bacteria were of two kinds, and both used sulfur in their system. Both needed sunlight as well, but they could not live in oxygenated waters. One was purple in color, and it was this species that lent the highly distinctive purple color to the lake's bottom water. Amid these were green bacteria, and these too were metabolizing sulfur.

But a third kind of bacteria was here as well, made up perhaps of several species, invisible to the naked eye. In their cells they produced hydrogen sulfide as a waste product of their metabolism. Only the thin layer of oxygen-laden water kept them from coming to the surface, where, if they could but get there, they would receive more light, grow faster, and release poison directly into the atmosphere.

We did not know it at the time, but in visiting this lake we had visited what would be recognized as the best modern analog of a hypothesized ancient ocean state that would be named a Canfield ocean, after geologist Don Canfield, who, with his mentor Robert Berner of Yale University, discovered evidence that Earth's oceans, long before the rise of animals, were chemically and biologically different from the oceans of today and were highly toxic, saturated by hydrogen sulfide. Our ocean, saturated with oxygen from top to bottom, is chemically far different, and far more benign, certainly to us animals, and even to most microbes. I had no idea at the time that those strange lakes would help answer that nagging question about the different fates of the ammonites and nautiloids, and certainly none that they would radically alter our understanding of mass extinctions. That understanding was still nearly three decades in the future. Until then, it was impacts all the way down.

Time was up for us two divers hanging on the anchor line. I remember taking one last look at the glorious reef around me. It was good, at this sublimely happy and peaceful moment, that I could not see into the future as well as I could see into the distant corners of this reef in such clear water, or even into the far reaches of time encapsulated in the sedimentary rocks I also studied.

I gave the thumbs-up to Mike, a man fated to drown almost exactly a year to the day after this dive, on a fine July morning in New Caledonia, and then have his lungs and heart popped by the remaining and expanding gases locked in his chest as I pulled him up from deep to shallow water in a rescue attempt. It would turn me away from studying the modern, and away from the sea, toward the landward study of darker things, the study of the mass extinctions themselves, for what better way to understand unexpected, unexplained death than to take its measure in its most sepulchral form?

And it was not just we that were doomed, each in our own way; even the ancient and vigorous Palauan reef around us was in its last years of life: In the early 1990s a large mass of warm, low-oxygen water would rise from the depths and kill all the corals of the Short Drop-Off, even those in the shallowest water. The lethal deepwater was very warm, that warmth having been generated by Earth's global warming. Today, like so many reefs around the world, the once thriving reef community at Palau's Short Drop-Off is a cemetery ultimately caused by anthropogenic carbon dioxide, a victim of what came to be known as coral bleaching, thanks to the washed-out colors it and other reefs would develop as they succumbed to water too warm. It would be one of the first shots of an oncoming greenhouse extinction, if my colleagues and I have correctly interpreted the clues from the past. The time for studying the nautiluses came and went, another decade passed, and with increasing heat the reefs began to die. *Something Wicked This Way Comes,* to steal a phrase from Ray Bradbury.

BY 2005, IT WAS PRETTY CLEAR THAT THE GEOLOGICAL AND BIOLOGICAL detectives knew what did *not* cause the Paleocene, Triassic, and most important, the immense Permian extinctions: asteroids from space. But eliminating impact as an extinction's cause (and at the same time, ripping the heart out of the now well-entrenched paradigm that impact had been the cause of most, if not all, mass extinctions) led to the very unsatisfactory state of not having the culprit in hand. If not impact, what? No one was going back to the twentieth-century saw about "slow climate change." How could slowly changing climate kill so many species? Likewise for flood basalts like the Palisades—even if they seemed the only viable alternative to impact, no one knew how they could kill anything. Although it was clear that the great flood basalts would have made Earth's air rich in carbon dioxide and thus would have led to rapid global warming, no one had been able to reconcile the effect—vast numbers of species killed—with the purported cause. Everyone assumed that if it became warmer, species would adapt by simply migrating pole-ward, for the increase in atmospheric heat from the volcanically produced carbon dioxide would have been on millennial or greater timescales, and such slow change—even if it was caused by enormous volcanoes—was just not a reasonable cause for a 90 percent death rate, such as Earth had suffered in the Permian extinction.

Some new ideas were needed. Happily, the time between the fall of the impact paradigm and the rise of its successor was not long.

Microbiologists studying the bacteria found in the jellyfish lakes of Palau and other similar kinds of anoxic lakes soon made a surprising discovery: that the varieties of bacteria in the waters left records of their presence in sediment. Microbiologists discovered that other organisms, and not just the peculiar microbes living in the Palauan lakes, left distinctive evidence of their presence too. Green plants using photosynthesis leave behind several distinct kinds of compounds, as do various kinds of microbes from other kinds of environments. A new kind of fossil was discovered.

Rather than looking for body fossils, microbiologists studying these strange, low-oxygen sites began to extract organic residues from the strata at the bottoms of their sampling sites, or even in the water itself, in search of *chemical* fossils, which are known as *biomarkers*. Other microbiologists, by studying modern organisms, figured out which biomarker came from which microbe. These biomarkers can serve as evidence of long-dead life forms that usually did not leave any skeletal fossils. Various kinds of microbes, for example, leave behind traces of the distinctive lipids, or fatty molecules, present in their cell membranes.

This biomarker research was first conducted on rocks predating the history of animals and plants, in part to determine when and under what conditions life first emerged on Earth. But within the past few years, scientists began sampling the mass-extinction boundaries. Using new kinds of mass spectrographs known as gas chromatography mass spectrometers, with skill and luck, investigators can tease out and identify what was there. Of greatest interest to the extinction detectives were the microbes living in water that was high in light, low in oxygen, and, to their surprise, high in hydrogen sulfide.

One such organism is a tiny species known as a photosynthetic purple bacterium. Today we can find such microbes in the Black Sea as well as lakes such as that in Palau. For energy they take up sulfur compounds—particularly hydrogen sulfide—and oxidize it. These microbes would be found only if other, more noxious characters were present as well—the bacteria that produce the hydrogen sulfide. Anyone who has taken freshman chemistry labs before the gas was banned from such teaching activities will remember how nasty and toxic the stuff is. Because of this extreme toxicity, most life avoids it. However, one large group of microbes is the exception to this. First near Australia, and then from numerous latest Permian-age strata from around the globe, it was confirmed that in case after case there was biomarker evidence of two kinds of microbes that inhabit water that must be low in oxygen but high in light and hydrogen sulfide. The light connection indicates that these were shallow waters, not the deep sea. It leads to a horrifying new view of the deep past, and to the tent pole to hold aloft a new paradigm for mass extinctions.

A team from Pennsylvania State University put the various pieces together. Lee Kump, one of the world's foremost experts on the chemistry of the oceans and especially its carbon cycle, along with his long-time colleague Mike Arthur (also of Penn State) and Alexander Pavlov (of the University of Colorado), published a bombshell paper in mid-2005 suggesting not only that there were great numbers of the nasty sulfur bacteria near the end of the Permian but also that the hydrogen sulfide that they produced was involved in the extinctions both on land and in the sea.

Only under unusual circumstances, such as those that exist in the Black Sea, do anoxic conditions below the surface permit a wide variety of oxygen-hating organisms to thrive in the water column. Those deep-dwelling anaerobic microbes churn out copious amounts of hydrogen sulfide, which dissolves into the seawater. As its concentration builds, the gas diffuses upward, where it encounters oxygen diffusing downward. So long as their balance remains undisturbed, the oxygenated and hydrogen sulfide–saturated waters remain separate, with a stable interface known as the chemocline. Typically the purple and green sulfur bacteria live in that chemocline, enjoying the supply of hydrogen sulfide from below and sunlight from above. Yet if oxygen levels drop in the oceans, conditions begin to favor the deep-sea anaerobic bacteria, which proliferate and produce greater quantities of hydrogen sulfide. In Kump and Arthur's models, if the deepwater hydrogen sulfide concentrations were to increase beyond some critical threshold, perhaps 200 parts per million, during such an interval of oceanic anoxia, then the chemocline separating the hydrogen sulfide–rich deepwater from oxygenated surface water could have moved up to the top abruptly.

So: If deepwater hydrogen sulfide concentrations increased beyond a critical threshold during oceanic anoxic intervals (times when the ocean bottom, and perhaps even its surface regions, lose oxygen), then the chemocline (such as those in the modern Black Sea) separating sulfur-rich deep waters from oxygenated surface waters could have risen abruptly to the ocean surface. The horrific result would be great bubbles of highly poisonous hydrogen sulfide gas rising into the atmosphere. This new entry into planetary killing can be referred to as the Kump hypothesis.

The proposal is relevant to more than just the end of the Permian; the same process may have occurred at other times in Earth's history and thus might have been the dominant cause of mass extinctions. Kump and his team did some rough calculations and were astounded to conclude that the amount of hydrogen sulfide gas entering the late Permian atmosphere would be more than 2,000 times greater than the small amount emitted by volcanoes today. Most likely, enough would have entered the atmosphere to be toxic. Moreover, the ozone shield, a layer that protects life from dangerous levels of ultraviolet rays, also would have been destroyed. Indeed, there is evidence that this happened at the end of the Permian period, for fossil spores from the extinction interval in Greenland sediments show evidence of being damaged by ultraviolet light, just the kind of damage expected from the loss of the ozone layer. Today we see various holes in the atmosphere, and under them, especially in the Antarctic, the biomass of phytoplankton rapidly decreases. (In fact, in late 2006, the hole over Antarctica was the largest ever observed.) If the base of the food chain is destroyed, it is not long until the organisms higher up suffer as well. (The complete loss of our ozone layer has even been invoked as a way to have caused a major mass extinction if Earth had been hit by particles from a nearby supernova, which also would have destroyed the ozone layer.)

Finally, the emergence of hydrogen sulfide from the seas would have coincided with an abrupt increase in both carbon dioxide and methane concentrations coming from the bottoms of the ocean that would have significantly amplified greenhouse warming from carbon dioxide pouring out of the eruptions—one of the largest in the history of the planet—that built the Siberian Traps. Hydrogen sulfide becomes more lethal as temperature rises, demonstrated in hideous lab experiments by physiologists long before Kump and his crew zeroed in on this poison as an extinction mechanism in which various animals and plants were exposed to hydrogen sulfide in closed chambers under conditions of ever-increasing temperature.

Kump's group undertook the difficult job of looking at the potential distribution of hydrogen sulfide emission around the globe. For this they used something called a global circulation model, or GCM. These models were originally developed to understand modern weather and climate patterns, but because the positions of the continents, as well as temperature, oxygen, and carbon dioxide levels in the atmosphere and oceans, are known for the critical period at the end of the Permian period and into the Triassic period, the method could be applied to the Permian. Lastly, Kump and his team looked for areas that would have seen high erosion rates for phosphorus-bearing minerals. Phosphorus is a prime component of fertilizer, and the sulfur microbes would have thrived if there had been an abundance of it; if oceanic phosphorus levels were observed to rapidly rise at the end of the Permian, the amount of hydrogen sulfide in the oceans and atmosphere would have jumped too. Because the level of the sea dropped at the end of the Permian, there would be vast regions with trapped phosphorus that had been underwater but that now eroded under rainfall and wind into the oceans, fertilizing them. Identifying them was tantamount to identifying the sources of hydrogen sulfide.

WASHINGTON, D.C., MARCH 2006

By 2006 the Kump hypothesis was enjoying ever-widening support as evidence in its favor kept coming in. Most important, geochemist Roger Summons of MIT found evidence for the presence of the hydrogen sulfide–producing microbes at the P-T boundary in nine places around the world. The toxic bloom was essentially global in extent. Questions remained, however, including whether there would have been enough hydrogen sulfide to actually kill things. Further, there was the question of whether all of this could be connected to the most salient evidence from the Paleocene thermal event, that the conveyer current system of the time had shifted to produce a warm, anoxic ocean bottom, or the main evidence from the Triassic mass extinction, that there was a series of mass extinctions, not just one, as evinced by the isotope record. The possibilities were exciting—it looked as if the evidence from the Paleocene, Permian, and Triassic extinctions could be forged into a new paradigm for mass extinction.

The Kump group presented new findings that added to their initial 2005 model at a large, NASA-sponsored astrobiology meeting held in Washington, D.C., in March 2006. It was the year's largest gathering of astrobiologists. Although much of the meeting dealt with more mainstream astrobiological topics, such as the new data from Mars and Titan showing that liquid of one kind or another had once been present on both bodies, or on the limits of extremophilic microbes on Earth, one afternoon was set aside for mass extinctions, for they were increasingly viewed as viable topics of astrobiology.

In a packed room, Kump and I presented back-to-back papers. In his, Kump reexamined the validity of his 2005 suggestion that it was hydrogen sulfide that actually killed things when the ocean states changed.

He showed a series of slides that in movie fashion showed that ancient Permian world. It was the oceans that were the critical element, and we all watched, fascinated, as the oceans became ever redder—the red chosen as the means to illustrate rising hydrogen sulfide levels. As they turned from pale pink to dark red, all the oceans were shown to be accomplices in the poisoning of the world. And perhaps most interesting of all, the overabundance of hydrogen sulfide did not happen only once but occurred over and over, as a succession of burps clustered around the time that the P-T boundary strata were being deposited around the world. Kump finished with the most ominous note. Not only did the model show where the hydrogen sulfide would emerge from the sea into the air, but he also showed new calculations that corroborated his earlier 2005 estimates of how much hydrogen sulfide would have eventually gone into the atmosphere. The results: There would have been more than enough to kill off most land life as the nasty stuff came out as bubbles. There would also be high levels of it dissolved into shallow seawater, where it would have been lethal in shallow marine settings as well, especially among shallow-water organisms that secreted calcium carbonate skeletons, such as corals, clams, brachiopods, and bryozoans, all invertebrate victims of the greatest extinction. Those organisms had already been teetering on the edge of extinction by the highly acidic seawater of the time, a product of the great volumes of carbon dioxide that entered seawater from the atmosphere. (Alarmingly this is occurring in our world in the Arctic Ocean, now so acidic that one group of mollusks, the pteropods, which are important in the food chain, are going extinct as their shells dissolve off their backs, as described by John Raven of the University of Dundee in Scotland and his colleagues in 2005.)

By the end of this session it was clear that the oceans were the key. But why would they change state?

A SHORT FERRY RIDE FROM SEATTLE LIES AN UPSCALE COMMUTER HAVEN called Bainbridge Island. Each morning thousands of suits take the 30-minute ride from suburbia to downtown offices and then rush back again at the end of the day. The tax base on the island, with its numerous waterfront and water-view houses is enormous, and why not—the view is sublime from the east side of the island looking at the magnificent cityscape of the downtown Seattle waterfront looming upward across two miles of Puget Sound. Even back in the Great Depression the wealthy valued this island, which became the site of the novel *Snow Falling on Cedars*, but not everyone who rides the boat to Bainbridge is a lawyer or plays golf. Some of the riders are students of various paleontology classes, for the southern tip of the island, on a closed country club, to be exact, is made up of 30-million-year-old sedimentary rocks that had been deposited on a fairly deep ocean bottom. Such outcrops are rare in the Seattle area, since the ice ages managed to dump untold tons of sand and gravel on the entire region repeatedly, covering the most useful teaching tools of a paleontologist, the rocks containing fossil life. The reason these outcrops are exposed is itself plenty ominous; the entire southern end of the island was thrown upward during the last mega earthquake that the region is prone to every 200 years or so. (The last was 200 years ago, and the tsunami wave generated by this monster quake crossed the ocean to devastate Japan, where its visitation was recorded in much art. Modern Seattle is a doomed city, each of its residents betting that a giant quake will not happen in our lifetimes. But what is life if not a gamble?)

The fossil-bearing rocks carry a salient message about the nature of the ocean back then: It was much like the ocean now—oxygenated from top to bottom—and we presume that this was maintained in some way by a conveyer current system analogous to that of today. By the Oligocene epoch of 30 million years ago, the world had cooled to something like its present state, after having been much warmer during the previous epoch, the Eocene. The ocean was warmer (as evidenced by a larger percentage of tropical snails and clams), but oxygen levels were the same at the bottom as at the top, and that has remained the case since. The abundance of life on the fossil bottom confirms that the oceans were animal-friendly from top to bottom, as does the fact that almost no bedding is visible in these rocks: On that ancient sea bottom a host of invertebrates managed to munch through the surface sediment to the extent that the original bedding was destroyed. We see this on the Bainbridge outcrops. No sedimentary bedding at all, just thick piles of well-sorted sediment rich in the shells of clams, snails, and other invertebrates. To a professional fossil finder, this kind of bottom is completely unlike the sea bottoms turned outcrops of older times.

The main drivers that created this mixed ocean were the extreme temperature differences that existed, and still exist, between the cool polar regions and the tropics. When there are warm surface areas and cold surface areas of the ocean, cold water spontaneously flows toward the warm, and vice versa. But more than surface currents accomplish this. Cold seawater is denser than warm water of the same chemistry and thus sinks. Saline water is denser than less saline water of the same temperature and also sinks. In the heat of the tropical sun, water rapidly evaporates, making the surface saltier and thus denser. In the Arctic, the melting of ice adds water to the sea, making it fresher. All of these factors create seawater bodies of different temperature and salinity that want to mix with others of different values, and in so doing produce conveyer currents throughout the world's oceans.

But this kind of ocean is a relatively new one. We have to go back only a slight way further in time to find a very different kind of ocean, one where the bottoms had very little oxygen. A longer ferry ride takes one north to the outermost island fringes of America, tucked into a larger archipelago of Canadian territory. There, too, rocky outcrops bear fossils, but both the rocks and fossils are very, very different from their younger Bainbridge counterparts. Here the black sea bottom beds show fine lamination and almost no fossils. The only remains are of surface-dwelling creatures of the time, fish, and chambered cephalopods such as nautiloids and ammonites. The layering was caused by the same sedimentary processes that were found on the younger Bainbridge Island beds, but the difference comes from the fact that unlike the Bainbridge beds, which were deposited on a well-oxygenated sea bottom, here there was an ocean bottom devoid of oxygen. The clues are numerous. Not only are there well-laminated beds but also numerous blebs of pyrite, or fool's gold, a sulfur-rich mineral that forms in the absence of oxygen. Sometime in the interval between the older fossil beds and the younger, the ocean radically changed.

The rock records and fossil records offer abundant testimony that this unmixed, or stratified, ocean, not the current mixed ocean, was far more common over most of geological time. The stratification involved temperature and salinity, and, for life, two far more important factors: dissolved oxygen and organic (reduced) carbon. They were characterized by an oxygenated surface layer, overlaying a much thicker water stratum with little or no oxygen. Encountering no oxygen at the sea bottom, the sediments accumulating on them became filled with black minerals colored by the abundance of sulfur within them; these sediments formed in a fashion similar to that responsible for the black layers found today on any beach when a clam digger gets below oxygenated sand and enters the thick black layer with its rotten-egg smell.

These black shales can be found all the way back to the dawn of life on Earth, at least 3.5 billion years ago. Paleontologists often love a good fossilized anoxic ocean bottom. Not for what lived there—there was little, and even less with shells available to fossilize—but because animals from the upper and still oxygenated layers fall to the bottom to be preserved, often in spectacular fashion. There are untold examples, the best being the life captured in the exquisite Burgess Shale, invertebrates and plants that fell onto a deep Cambrian anoxic bottom that preserved even their soft parts as well as the more commonly fossilized skeletons. Nice! And how about *Archaeopteryx* and much else from the Solnhofen limestone of Jurassic Germany, those early birds whose bodies fell onto an anoxic bottom, a place bereft of the scavengers that usually feast on such fowl. No scavengers ruffled those first feathers spread out on the bottom sediment in sprawling death. But for other kinds of life, like us animals, the low-oxygen conditions are highly inimical.

The stratified oceans can themselves be subdivided into two kinds. When oxygen levels at the bottom are just low, there may still be a few animals here and there, or maybe not. But the most common organisms by far on these bottoms are microbes. Even a tiny bit of oxygen, too little to support animals, is enough to maintain one kind of microbe, although it plays no part in the affairs of us animals. But when oxygen levels really reach bottom, a very different kind of microbial kingdom takes over, one dominated by bacteria that use sulfur as foodstuff. These are the nasty forms that make the poisonous hydrogen sulfide. At times when they have been present in abundance—and this can only be ascertained by finding their characteristic biomarkers in the organic fraction of the rocks making up these ancient sea bottoms—we say that the ocean containing them was a Canfield ocean.

So toxic were Canfield oceans that they might have reduced animal life, or even inhibited its first evolution for millions of years during the long-age Precambrian era, which includes the time from life's origin to less than 600 million years ago. There seem to be two reasons for this. First is the obvious toxicity of the hydrogen sulfide, but just as important may have been the microbes' inhibition of nitrogen formation in compounds useful for plant life. While many kinds of microbes can "fix" biologically useless nitrogen, an essential element for life, into compounds that are biologically useful, the eukaryotes—plants, animals, fungi, and a variety of other groups—cannot do this trick and so depend on microbes to do the job for them. Enter the Canfield ocean's gang of sulfur bacteria, and little nitrogen becomes available, because this kind of bacteria couldn't care less about nitrogen, and also inhibits other microbes from supplying it. A nitrogen-poor ocean would have been an ocean literally in need of fertilizer and not getting it. It would have been just like a soil from which all the nitrogen has been leached—only a small amount of plant life will grow. Nasty place, that Canfield ocean. Perhaps if a mixed ocean turned into a Canfield ocean, a great mass extinction would soon follow. Is there any evidence that these Canfield oceans, which we know existed in the time before animals, made their destructive returns, like a bad plague, in the time of animals as well?

Yes. The most important of the mass extinctions was clearly at the P-T extinction, and it was indeed a time of a Canfield ocean, an identification made in 2005 when a team led by biogeochemist Kliti Grice of the Curtin University of Technology in Perth, Australia, published a seminal paper in *Science* on research that demonstrated that the oceans at the end of the Permian period showed biomarkers of the microbes that would be expected in a Canfield ocean. The second, the T-J extinction, is just now being examined, but already beds from the Alps have shown the presence of isorenieratane, the biomarker characteristic of the purple and green sulfur photosynthesizing bacteria, forms that can live only in seas shallow enough for light to penetrate that are also low in oxygen and high in hydrogen sulfide concentrations. We do know that the last few million years of the Triassic period and the first few million years of the Jurassic period were characterized by a series of isotopic perturbations that coincided with pulses of anoxia in the sea, and both of these strongly suggest that a series of short-lived Canfield oceans led to the sequential series of mass extinctions that, combined, we call the T-J mass extinction.

Three different ocean states—the mixed ocean, and two kinds of unmixed ocean, the anoxic and Canfield oceans. How and when does one become one of the others? Here is where the conveyer current systems come in. They seem to be the gatekeepers for determining which ocean type will be present. And it may not be the *presence* of any of these oceans that causes distress to life but the *change* from one state to another.

THE SOURCE OF THE MASS EXTINCTIONS WAS A CHANGE IN THE LOCATION at which bottom waters are formed. Near the end of the Paleocene epoch, the source of our Earth's deepwater shifted from the high latitudes to lower latitudes, and the kind of water making it to the ocean bottoms was different as well: It changed from cold, oxygenated water to warm water containing less oxygen. The result of this was the extinction of deepwater organisms that Jim Kennett and Lowell Stott were investigating. The cause of the Paleocene event is thus linked to a changeover of the conveyer belt system. What about the biggest of all extinctions, the Permian? It turns out that for it, too, a changed conveyer current holds the smoking gun.

In 2005, climatologists Jeffrey T. Kiehl and Christine A. Shields of the Climate Change Research Section at the National Center for Atmospheric Research used a global circulation climate model to look at the Permian world. Kiehl and Shields wanted to know if Permian ocean circulation patterns were disrupted at the time of the extinction. When they plugged in the known positions of the continents and inputted a warmed world as well, their modeled Permian world showed a shift in the positions of its conveyer belt currents. They proposed that sudden global warming caused a change in ocean state.

Oceanic currents play a huge role in current climate and global temperature. Today, whether the conveyer current system in the north Atlantic Ocean runs seems to be controlled by the amount of ice cover on Earth, and in a complicated fashion (no weather is ever simple, alas) by the nature of tropical warming or cooling. By the end of the Permian period, Earth appears to have had no ice—the ice caps had all melted away from their early Permian maximum. (The early Permian period of 300 million to about 270 million years ago was so globally cold that there were vast continental glaciers resembling those of our own recent Ice Age. By the late Permian period, some 260 million to 250 million years ago, however, they were either gone or going fast, according to the geological evidence from these time intervals.) The conveyer current does not shut down in the absence of ice. Rather, it shifts the positions of its starting and ending points (where water either comes up from depth or dives down to ocean bottoms). That shift may have been crucial in the mass death that followed.

Because the continents were in such different positions at that time, models we use today to understand ocean current systems are still crude for the Permian oceans, and they have much less precision than those we can make for the modern world. Nevertheless, it seems fairly clear that by the end of the Permian period, ocean circulation had changed so that the deep ocean bottoms filled with great volumes of warm, virtually oxygen-free seawater. This seems like the same thing that happened at the end of the Paleocene epoch but at a vastly increased scale, and with vastly more destructive results. The Permian bottom waters were warmer than those of the Paleocene and much less oxygenated. The stage was set and needed but one more trigger, and it seems both had the same trigger—a short-term but massive infusion of greenhouse gases into the atmosphere changed the nature of the oceans. In the Paleocene epoch the source of that carbon dioxide was volcanism in the Atlantic Ocean region, whereas at the end of the Permian period the initial source of the heat was emission of vast volumes of carbon dioxide from the spectacular lava outpourings, perhaps one million cubic miles in volume, that today cover some 800,000 square miles and that might have covered nearly three million square miles when formed into what is known as the Siberian Traps. (Why igneous geologists call stacked-up piles of lava "traps" is beyond me.)

Now, it seems, events at the end of the Permian period can be related to changes in oceanography as well, with the addition of a kill mechanism from hydrogen sulfide that was microbially produced.

The main difference between the two events seems to be that the Permian event showed far more upwelling of poisonous bottom waters. In the case of the Paleocene event, some deep, near-shore basin underwent a change from oxygenated to less oxygenated, even into the shallows, after upwelling of the deep, warm water, in a manner happening today to the Gulf of Mexico, Gulf of California, and Puget Sound, among other such places. Deep, warm water also upwelled at the end of the Permian period, but it did so over a far greater area of the globe—virtually every shallow-water area (rather than just a few, as at the end of the Paleocene epoch) became filled with warm water without oxygen, even at the surface. And the Permian deepwater brought up poison not seen at the end of the Paleocene—it was rich in carbon dioxide and methane, which seems to have moved out of the solution in seawater and into the atmosphere as potent greenhouse gases (causing even faster planetary warming) as well as the deadly hydrogen sulfide gas, which, if it occurred a the end of the Paleocene, did so only at low concentrations.

IN A 1997 BOOK, *MASS EXTINCTIONS AND THEIR AFTERMATH,* ANTHONY Hallam of the University of Birmingham in England and Paul Wignall of the University of Leeds compiled what was known about all the mass extinctions in an excellent volume. At the time the impact hypothesis still held sway. Nevertheless, their data show that of the 14 mass extinctions they recognized, 12 of them were characterized by poorly oxygenated oceans, which they thought must have been a major part of the cause of the extinctions. There is no proposal about how the oceans got that way, however. With the models above, we now have a mechanism: perturbation or even stoppage of the thermohaline, conveyer current systems. It is time to stop looking at the "kill mechanisms"—low oxygen, heat, and perhaps excess hydrogen sulfide gas in water and air—and start looking at the driver of these changes, the atmosphere itself.

CHAPTER 6

The Driver of Extinction

The flight into Walvis Bay, Namibia, is one of wonders. You cross the great expanse of South Africa and its Great Karoo Desert, and then the far drier Kalahari, finally seeing the coastline of the southern Atlantic Ocean along this southern African nation. The whole coast is often obscured in thick fog, a cool coastal blanket keeping the scorching desert heat a few miles to the east at bay. The bay itself hosts untold numbers of bright, wild flamingos on its intertidal mud and wading in the frigid sea. The ocean there is so cold that no human dares enter without thick thermal protection. It seems strange to have such a cold sea, penguins and all, next to one of the hottest places on Earth, but we weren't there for this modern coast, and a long drive finally finished at a rocky outcrop where several ancient oceans and climates sit stacked one atop another, a reminder that there was more than one kind of ocean in the past, just as there will be a new ocean to come in our immediate future, and if that future is too immediate, God help us all.

At the base of these rocks, deposited some 700 million years ago on an equatorial sea bottom, are thick mudstones filled with larger rocky cobbles of many different varieties. The two rock kinds seem incompatible: Normally, large rocks are not found in mudstones deposited well offshore. In our world such rocks form only in a particular environment—under the path of floating icebergs, themselves the remains of glaciers calving into the sea in some cold place. Icebergs are floating rock collections. As glaciers move across the wilderness, they scoop up pieces of country rock, grind them round beneath the moving ice, and freeze them into an icy place; the embedded rock floats away once the ice becomes an iceberg. On sunnier days the icebergs begin to melt, and as they do so, they drop their cobbles onto the deep ocean below, creating rocks identical to those we visited there in Namibia on a burning outcrop that made a glacier seem like a joyous possibility of coolness.

The deposits containing the ancient glacial drop stones were thick but not endless. At the top of the outcrop a very different kind of rock appears: thick, layered limestones bearing signs of bacterial life. These are known as stromatolites, and again somewhat similar rocks can be seen forming today. But this time we traveled not to the frigid Arctic or Antarctic to see these rocks form but to the steaming reef regions of the tropical, equatorial seas. So—a mystery—rocks suggesting cold, overlain suddenly by rocks showing warm.

We saw there evidence of what has been called a snowball Earth, an episode of cold so intense that the oceans may have frozen from pole to pole, only to be rapidly melted when some threshold level of volcanically produced carbon dioxide caused a warming over the icy oceans sufficient to melt all away and build warm-water limestones where only glaciers and sea ice once stood. We saw as well a changeover from one kind of ocean to another, the first an ocean lacking the mixing currents and wave energy that sends oxygen to deeper depths. And then all that changed, becoming well mixed with oxygen.

The thawing of the world underscores the power of a force that seems ridiculously small to have such great effect on a world. It was the rapid rise in greenhouse gases that ended the snowballs, and the most important of the gases that accomplished these thaws are measured in parts per million. At such low levels in the world's atmosphere, even small additions or subtractions can have a great effect on the temperature of the world—an increase of a few hundred parts per million can cause the world to heat radically, while equivalent drops will cool the planet. And because of the small amounts of greenhouse gases needed, the climate can change very quickly, with what are increasingly formalized as rapid climate changes.

Because of the importance of greenhouse gases in controlling climate, discerning how much carbon dioxide and methane there has been in Earth's atmosphere over time has attracted the efforts of many atmospheric scientists. Unfortunately no direct measure in the rock record gives a determination of actual levels in the deep past (beyond the recent past, that is; direct readings of carbon dioxide trapped in ice in continental ice sheets can be made). Nevertheless, scientists have come up with two very different but clever ways of estimating past levels of carbon dioxide. One comes from mathematical modeling, the other from measurements on fossil leaves.

The best of the methods is a computer program developed by the great Robert Berner of Yale University. His program depends on inputs ranging from estimates of the rate of sediment burial to direct measurement of carbon isotope values from carbonate rocks. Known as GEOCARB, the model shows the major trends in carbon dioxide through time, while a second such model known as GEO-CARBSULF allows estimates of ancient oxygen levels through time.

Combined, these results have given us a new, and in many ways largely unexpected, view of how much these two gases have varied through time.

That levels of atmospheric carbon dioxide must have varied through time became evident after geologists discovered times in Earth's history when much of Earth was tropical, and other times when there were glaciations on larger scale than the recent Pleistocene epoch glacial events. Although there were several possible causes, such as variation in solar heating over time or changes in heating from the interior of Earth, detailed research into both eventually ruled them out, leaving greenhouse gases as the major suspect for having caused climate shifts. The study of ice cores from glaciers formed in the Pleistocene epoch finally demonstrated that carbon dioxide values can and did vary, and not just in the long term but over astonishingly short time intervals, some as short as a decade. Although none of the various models such as GEOCARB attempting to calculate carbon dioxide and oxygen levels over the last 500 million years has such precision, they can discern longer-term (greater than a million years) changes (Figure 6.1).

The carbon dioxide curve is striking. Compared with today, it was high for much of the Paleozoic era, but as oxygen began its climb some 375 million years ago, the levels of carbon dioxide plummeted and only rose again sometime into the Mesozoic era; the gas was plentiful through much of the era, culminating in a maximum in the late Jurassic period, of about 150 million years ago, and then declining throughout the Cenozoic era, coming to a minimum level today. But as we shall see, even the last 200 years have produced an upswing of carbon dioxide that is of too short a duration to be visible on this long-term graph.

The long-term decline in carbon dioxide over the past 100 million years is both interesting and misleading. The gradual reduction is due to the slow enlargement of the continents and to the increased

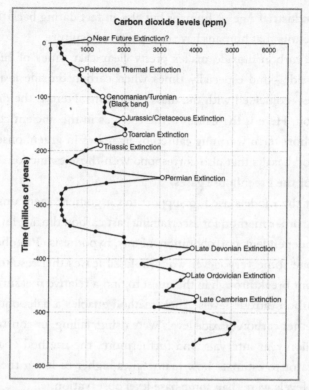

FIGURE 6.1

Carbon dioxide through time, as computed with GEOCARB, a computer program for estimating past levels of carbon dioxide. Each of the open bullets designates a mass extinction, and they can be seen to correspond either with high or sharply increasing carbon dioxide levels.

amount of carbon locked up in vast mineral deposits, which erode and liberate carbon at a slower rate than others of their composition are formed. Ultimately, the long-term drop will spell doom for our Earth as a habitable planet, as Don Brownlee and I explained in our 2003 book, *The Life and Death of Planet Earth*. But that eventuality is far, far in the future, and the reduction in atmospheric carbon dioxide has halted and reversed with a vengeance over not only the past centuries of the Industrial Age of humans but also in fact dating back through the millennia that humans have engaged in agriculture.

The carbon dioxide makes pretty clear that times of high carbon dioxide—and especially times when carbon dioxide levels rapidly rose—coincided with the mass extinctions. Here is the driver of extinction. Here is the cause of the changes in the ancient conveyer belts—short-term warming caused by increases in greenhouse gases. The flood basalts that also correspond with those extinctions are the source of the greenhouse gases.

Lest the models used to support this argument seem unsatisfactory, the other method for ascertaining past carbon dioxide levels provides independent corroboration of my hypothesis. Paleobotanists have done some very clever work on fossil leaves that resulted in an important breakthrough in the quest to find a relative measure of ancient carbon dioxide levels. Their method enables a paleobotanist to say whether carbon dioxide levels were rising, falling, or constant during million-year intervals, and furthermore, the method enables an investigator to estimate how many times higher or lower the carbon dioxide levels were than some base-level observation.

The measure turns out to be both clever and simple, as is so often the case with wonderful breakthroughs. Botanists looking at modern plant leaves had done experiments whereby they grew plant species in closed systems where the amount of carbon dioxide could be raised or lowered relative to the level found in our atmosphere (about 360 parts per million when these experiments were first conducted). Plants, it turns out, are highly sensitive to carbon dioxide levels, because the carbon dioxide in the atmosphere serves as their source for carbon, the major building block of life. They acquire this mainly through tiny portals in their leaves called stomata; the stomata allow carbon dioxide in and water out. When grown in high levels of carbon dioxide, the plants produced a small number of stomata, as just a few sufficed with the gas plentiful; when grown in low levels, the opposite was true. Such a clear result delighted the experimenters, which included a colleague of Berner's named David Beerling, now teaching at the University of Sheffield in England. Leaf stomata are readily observable on most well-preserved fossil leaves, and when the investigators turned to the fossil record, the results confirmed Berner's model results.

LET US BRING THIS ALL TOGETHER. IT IS HERE PROPOSED THAT EACH OF the greenhouse extinctions had a similar cause, and here we can summarize the sequential steps.

First, the world warms over short intervals of time because of a sudden increase of carbon dioxide and methane, caused initially by the formation of vast volcanic provinces called flood basalts. The warmer world affects the ocean circulation systems and disrupts the position of the conveyer currents. Bottom waters begin to have warm, low-oxygen water dumped into them. Warming continues, and the decrease of equator-to-pole temperature differences reduces ocean winds and surface currents to a near standstill. Mixing of oxygenated surface waters with the deeper, and volumetrically increasing, low-oxygen bottom waters decreases, causing ever-shallower water to change from oxygenated to anoxic. Finally, the bottom water is at depths where light can penetrate, and the combination of low oxygen and light allows green sulfur bacteria to expand in numbers and fill the low-oxygen shallows. They live amid other bacteria that produce toxic amounts of hydrogen sulfide, and the flux of this gas into the atmosphere is as much as 2,000 times what it is today. The gas rises into the high atmosphere, where it breaks down the ozone layer, and the subsequent increase in ultraviolet radiation from the sun kills much of the photosynthetic green plant phytoplankton. On its way up into the sky, the hydrogen sulfide also kills some plant and animal life, and the combination of high heat and hydrogen sulfide creates a mass extinction on land. These are the greenhouse extinctions.

The sequence of events outlined above can be considered a combined hypothesis for the cause of greenhouse extinctions and can be named the conveyer disruption hypothesis. There was obviously variability in each extinction, but if the extinctions are examined in a fashion similar to how taxonomists classify living organisms as a species, it seems quite clear that the mass extinctions considered here as greenhouse extinctions are a different beast than the K-T, our now sole known impact extinction.

What would Earth be like in the midst of such an event? Let us crank up a hypothetical time machine and visit one. We have a lot of choices of where to go, back in time: the mass extinctions ending the Cambrian, some 490 million years ago; the late Ordovician mass extinction, some 450 million years ago; several late Devonian mass extinctions, around 360 million years ago; the Permian mass extinction(s), ranging from 253 million to about 247 million years ago; the Triassic mass extinctions, ranging from 205 million to 199 million years ago; the Toarcian mass extinction, some 190 million years ago; the Jurassic–Cretaceous mass extinction, some 144 million years ago; Cenomanian–Turonian mass extinction, some 93 million years ago; and the Paleocene thermal event, some 55 million years ago. All are united by cause, increased carbon dioxide in the atmosphere, leading to change in ocean currents, and eventual anoxia. Just because we get to see some dinosaurs, let's go back to near the end of the Triassic period, to the site in Nevada that begins this book:

No wind in the 120-degree morning heat, and no trees for shade. There is some vegetation, but it is low, stunted, parched. Of other life, there seems little. A scorpion, a spider, winged flies, and among the roots of the desert vegetation we see the burrows of some sort of small animals—the first mammals, perhaps. The largest creatures anywhere in the landscape are slim, bipedal dinosaurs, of a man's height at most, but they are almost vanishingly rare, and scrawny, obviously starving. The land is a desert in its heat and aridity, but a duneless desert, for there is no wind to build the iconic structures of our Saharas and Kalaharis. The land is hot barrenness.

Yet as sepulchral as the land is, it is the sea itself that is most frightening. Waves slowly lap on the quiet shore, slow-motion waves with the consistency of gelatin. Most of the shoreline is encrusted with rotting organic matter, silk-like swaths of bacterial slick now putrefying under the blazing sun, while in the nearby shallows mounds of similar mats can be seen growing up toward the sea's surface; they are stromatolites. When animals finally appeared, the stromatolites largely disappeared, eaten out of existence by the new, multiplying, and mobile herbivores. But now these bacterial mats are back, outgrowing the few animal mouths that might still graze on them.

Finally, we look out on the surface of the great sea itself, and as far as the eye can see there is a mirrored flatness, an ocean without whitecaps. Yet that is not the biggest surprise. From shore to the horizon, there is but an unending purple color—a vast, flat, oily purple, not looking at all like water, not looking like anything of our world. No fish break its surface, no birds or any other kind of flying creatures dip down looking for food. The purple color comes from vast concentrations of floating bacteria, for the oceans of Earth have all become covered with a hundred-foot-thick veneer of purple and green bacterial soup.

At last there is motion on the sea, yet it is not life, but anti-life. Not far from the fetid shore, a large bubble of gas belches from the viscous, oil slick–like surface, and then several more of varying sizes bubble up and noisily pop. The gas emanating from the bubbles is not air, or even methane, the gas that bubbles up from the bottom of swamps—it is hydrogen sulfide, produced by green sulfur bacteria growing amid their purple cousins. There is one final surprise. We look upward, to the sky. High, vastly high overhead there are thin clouds, clouds existing at an altitude far in excess of the highest clouds found on our Earth. They exist in a place that changes the very color of the sky itself: We are under a pale green sky, and it has the smell of death and poison. We have gone to the Nevada of 200 million years ago only to arrive under the transparent atmospheric glass of a greenhouse extinction event, and it is poison, heat, and mass extinction that are found in this greenhouse.

THIS SHOULD THUS BE THE END OF THE BOOK. IMPACT ONLY RARELY causes mass extinction. But it is the realization by an increasing number of us of just what *did* cause the other mass extinctions that should make every citizen stand up. The beauty of dinosaur stories is that no matter how ferocious or dangerous they are in the movies, that is all that they are: in the movies. Here, however, we have a process that is very real—mass extinction—and the understanding that conditions on Earth now in some ways seem similar to the causes of the mass extinctions of the past. Carbon dioxide is carbon dioxide, whether it comes from a smoking volcano or a smoking car. The question thus becomes one of whether the rate of carbon dioxide increase in our world is on par with the rate during those times when greenhouse extinctions occurred. Just how much danger are we in, anyway? To answer that, we need to look at the "natural" rates of carbon dioxide formation as well as the human rates and see if our modern world is so different from the deep past that it renders the current, rising carbon dioxide levels less dangerous than they were so many times in the past.

We have thus come to a point where the past meets present. What do modern events look like when examined through past-tinted glasses?

CHAPTER 7

Bridging the Deep and Near Past

I magine a giant canyon separating two groups of humans. It is so wide in places that most do not know that another group is on the other side; some do not even seem to care that there *is* another side. And to those who do look across, the canyon is seemingly unbridgeable. So they shrug, wave once in a while, and return to their labors. They are two groups of dedicated scientists, each working feverishly on scientific problems endemic to their side of the chasm.

On one side is the vast army trying to figure out present-day climate. There are many varieties of work involved. Some use models, some study ice-core records, some look at peat bogs; in fact, there are so many kinds of research going on that no single worker can keep track of it all, so that even on one side of the vast chasm there are smaller ravines separating the many students of modern climate, ravines that effectively keep specific groups from talking to, or, more important, working with others from a slightly different discipline.

On the other side of the vast canyon is another group of scientists, also working away with prodigious effort. These are the scientists interested in understanding climate change in the deep past, the kind of climate change that seemingly led not only to the mass extinctions hundreds of millions of years ago but also to the longer-term trends that warmed the Mesozoic era, for instance, and then cooled the ensuing Cenozoic era. Like the group on the other side, these workers are also so diverse in interests that they talk little with others on their side of the canyon.

Most on each side do know of the existence of the other. But they are so caught up in their own work, their noses so close to their particular problem, that there is only the occasional wave to the other side. They intuitively know that they would have much to talk about if, somehow, a bridge could be built across the chasm. But that has not happened, and the real surprise is how little effort has really taken place to build that bridge. Perhaps that is about to change, for increasingly, workers on each side are seeing that their individual efforts are constructing pieces that could fit with others into an elaborate puzzle. Neither side, even if all their pieces are united, has enough of these pieces to make out the identity of the larger picture itself. But if all the pieces were combined, the identity of the puzzle's illustration could surely be solved.

What might the picture be? Perhaps it is a view of Earth in one to two centuries, or several pictures based on what our society does over the next few decades. Perhaps in that picture there is only a warmer world, a pleasant place with less snow. But then again, perhaps that pleasant dreamlike view is a chimera, and the true picture is one of raging storms, failing crops, and human famine leading inevitably to war for more fertile territory, a picture of human deaths numbering in the millions or higher, a stark portrait of a world entering a long, slow slide into mass extinction.

There may be one more picture as well: clues as to how to avoid the nightmare.

If we think of the gap as a break between disciplines, we can even define it in terms of a particular time. The break in interest, models, and modes of study between the two sides coincides with onset of the most recent ice age. From the initiation of that interval to the present time is one of the groups, while the other group studies the time interval from the Pliocene epoch on back to the earliest history of Earth. The reasons for this are many, but perhaps most significant are the radically different climates of the Ice Age compared with immediately before.

If a bridge were built, a really sturdy bridge allowing many to cross and intermingle at the same time, rather than one allowing rare, solo crossings of high specificity, what might the conversations be about? Could the various visitors to the respective foreign side even speak the same language as their new hosts? What would they have in common, and if prioritizing, what projects might they most fruitfully undertake together? There would be many, of course, but I know how I would vote when on the other side, the side new to me, the side populated by climatologists of the now as well as those looking back into the ice ages of the last 2.5 million years. I would vote for those who now work on the oceanic thermohaline conveyer currents to examine the evidence of those who study the past mass extinctions.

LEAVE IT TO HOLLYWOOD TO MAKE A MOCKERY OF A SERIOUS SUBJECT, changing the laws of physics in search of a better story line. So it was with the laughable movie *The Day After Tomorrow*, a fable about rapid climate change that suddenly plunged the civilized world into a killer winter at the drop of a hat, or some such really fast thing. The tragedy here is that something far more ominous—rapid climate change occurring not in days but in the actually far more catastrophic interval of years—is trivialized in a movie. We have seen Hollywood do this before, of course, most recently and famously with the very serious threat of asteroid impact on Earth. But we modern humans have now lived through a long period of climatic stability. Of course there are variations that seem epochal—the appearance and disappearance of El Niño currents, for instance, of decadal droughts that seem to portend the end of a climate as some given region knows it—only to prove minor and give way to an eventual restoration of what had been considered the norm. But if we go back in time at a millennial pace, it is not too long a journey until we reach ancient times when rapid climate change—real change—was the norm, and it was something no silly movie could prepare us for.

That rapid climate change was not only possible but the norm for the time we call the ice ages is a relatively new discovery. One of its discoverers was a man who dedicated his life to improving the way science can date the past. His name is Minze Stuiver, and before his recent retirement he was one of the biggest fish in a not insubstantial pond, the Quaternary Research Institute at the University of Washington, a group of scholars who specialize in the last 12,000 years or so, the time since the last of the great continental ice sheets of the Ice Age finally melted back into high mountain keeps. Stuiver, with his European accent of somewhat Germanic ancestry and tall military bearing, spent the 1960s through the 1990s improving techniques of carbon-14 dating in a gigantic, windowless laboratory buried deep underground to avoid cosmic rays and other particle riffraff that could affect the precise measurements necessary to date very old material. Because he had the best lab on Earth to date objects less than 100,000 years old, it was natural that a new kind of samples, coming from a novel attempt to core thick ice deposits of unknown antiquity in Greenland and later Antarctica, would be sent to his lab.

For thousands of years, Greenland has been a cold, forbidding place of howling winds amid snow and ice. Greenland carries one of Earth's greatest reserves of fresh water locked up as ice, ice that has been deposited, year after year, for more than 2,000 millennia. It is from this ice that scientists such as Stuiver uncovered one of the most startling discoveries of the twentieth century. On these cold continents and islands, these ancient piles of ice were found to be made up of individual ice layers, laid down year by year, so that the cores, if studied in enough detail, could yield a yearly record of ice accumulation spanning back hundreds of thousand of years. Stuiver and others were subjecting these cores to measurements with mass spectroscopes that allowed them to discover both the temperature at which the core was formed and the amount of carbon dioxide in the atmosphere at the time a particular level of the ice sheet that would eventually be cored was formed. Thus he was privy to an unprecedented, ultimately high-resolution record of the past coming from the ice record of Greenland.

After two decades of patient collection, followed by interminable isotopic analyses of the ancient fossil ice, scientists from Europe and America such as Stuiver finally completed their analyses of Greenland ice-core samples dating back 200,000 years. They had expected to find a record of climate stability interspersed with epochs of temperature change, each coinciding with the advance and retreat of the great Ice Age glaciers. They found nothing of the sort. The emotionless, unblinking numbers emerging from the great mass spectroscopes across the world showed that geologically recent fluctuations of Earth's climate have been far more severe and have occurred much more quickly than any scientist had previously postulated—up until 10,000 years ago, that is. This new discovery allows an entirely new interpretation of the rise of human civilization and certainly shows that the present-day weather—one of the prime bases for the concept of uniformitarianism—is in fact very aberrant. We are currently in a state of calm, a period lasting 10,000 years. Prior to that, things were anything but.

Stuiver and many others working on the ice-core record showed that 200,000 and 10,000 years ago, the average global temperature had changed as much as 18 degrees Fahrenheit in a few decades. The current *average* global temperate is 59 degrees Fahrenheit. Imagine that it suddenly shot to 75 degrees Fahrenheit or dropped to 40 degrees Fahrenheit, in a century or less. We have no experience of such a world and what it would be like; such sudden perturbations in temperature would enormously alter the atmospheric circulation patterns, the great gyres that redistribute Earth's heat. At a minimum, such sudden changes would create catastrophic storms of unbelievable magnitude and fury. Yet such changes were common until 10,000 years ago. Imagine a world where devastating storms like 2005's Hurricane Katrina lash the continents not once a decade or century but several times each year, every year. Imagine a world where tropical belts are suddenly blanketed by snow each year. This was our world until 10,000 years ago. According to the new studies from Greenland, a miracle then happened—the sudden shifts in the weather disappeared.

Very quickly after the start of this calm, we as a species began to build villages and then cities; we learned to smelt metal and conquer nature. And most important, we learned how to tame crops and domesticate animals. Human population numbers began to soar as larger mammals underwent wholesale extinction. But there were still a few climate bumps in the road for both humans and animals, for the climate has had a history of rapid change. How rapid were these changes? The answer was unsettling indeed. The ice-core work indicated that a global temperature change of 10 degrees Fahrenheit could take place in as little as 10 years.

There remain many mysteries, however. For example, new ice-core recovery from Antarctica seems to suggest that the large tem-

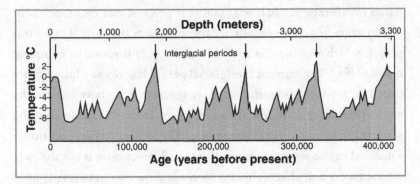

FIGURE 7.1

Variation of average temperature in last 400,000 years

perature swings observed in Greenland ice did not affect Southern Hemisphere ice. However, the work on the Antarctic ice cores is just beginning, and these results could still be from inadequate sampling, although that conclusion grows more unlikely with each passing month (Figure 7.1).

So, now a rate for climate change was known. But what could cause such rapid changes, or the even bigger changes that brought about the advance or retreat of whole continental glaciers? Climatologists have long theorized that climate change observed over the past million and a half years, the alteration between long periods of very cold climate with growing ice sheets and dropping sea level alternating with shorter times of warmth, were the result of the orbital changes first articulated by Malutin Milankovich. Until the ice cores became available, with their unprecedented resolution in discerning climate through recent time, the changes were thought to have been slow. But with that resolution a newer view became apparent.

The ice-core records and other sources of climate information, such as deep-sea paleontological and isotopic records indicate that over the past 800,000 years, the warmer interglacial periods have lasted on average for half a 22,000-year precessional cycle, or about 11,000 years. (Each precessional cycle measures the time it takes our planet, which wobbles on its axis like a top as it spins, to complete one wobble.) The current interglacial period has already lasted more than 11,000 years, and some records suggest that we have been in the warm period for as much as 14,000 years. Does this mean that the glaciers are advancing at this moment? The answer to that question is a decided *no*, for several reasons. First of all, precession is not the only orbital aspect that affects climate. Records show that between 450,000 and 350,000 years ago there was an interglacial stage that lasted much longer than 11,000 years. This interglacial period was coincident with a time when orbital eccentricity was at a minimum—that is, the orbit was closer to being a circle than an elongated ellipse. Just such a pattern of minimal orbital eccentricity is under way at this time, suggesting that the present interglacial period could continue for thousands, to perhaps a few tens of thousands of years into the future—or it could end at any time.

By using details from the past we can even extrapolate further into the future. One such prediction was recently made by R.C.L. Wilson, Stephen Drury, and Jenny Chapman in their masterful book *The Great Ice Age*, published in 2000, in which they predict that the interglacial period should end within a few more thousands of years at most, to be followed by a pattern that has been present for the past two million years or more: a drop of global temperature by as much as 10 degrees Celsius (or nearly 20 degrees Fahrenheit) for the next 80,000 years, and in all of that time our planet would never experience temperatures approaching those of the present day. But that would be the case if humans were not around. Now, this prediction seems laughable.

Is there any way to deduce the ultimate cause of these climate changes? Once again, as we saw for the deep past, it seems that changes in the conveyer current system might have been involved. Evidence from the Permian and Paleocene events (and, we suspect, from other of the more minor mass extinctions as well) indicates the cause of most past mass extinctions in deep time: that they might have started by short-term global warming, causing a perturbation in the conveyer current systems presumed to have been present in the oceans through time. That leads us to two important questions: What is the relationship between an Ice Age conveyer system and rapid climate change? Can humankind, through the release of greenhouse gases, cause a change to the modern conveyer? Let us try to answer these two questions.

First, let us look at the nature of the conveyer over the last 2.5 million years. Many scientists now believe that it changed from time to time. Not by changing position, but by turning on and off. The evidence for this comes from the ice cores profiled above, as well as from the study of rock cores extracted from the bottom of the Atlantic Ocean.

The ice-core evidence that there have been very rapid changes in climate, later interpreted as having been caused by the shutdown of the conveyer belt current, was first made in the late 1960s by Danish geochemist Willi Dansgaard, who came to this startling conclusion when analyzing the first primitive attempts to interpret early ice cores taken from Greenland. Later, Swiss climatologist Hans Oeschger examined better cores, and both eventually found that the ice cores seemed to suggest that for most of the last 100,000 years, an abruptly changing climate was the rule, not the exception. They found evidence that the Earth slowly underwent a slight and gradual cooling over centuries but then underwent an abrupt cooling in decades or less, and then stayed really cold for a thousand years or more. Then there was a second abrupt change to warm climate, and the cycle renewed, an entire single cycle lasting about 1,500 years; the pattern has come to be called the Dansgaard–Oeschger cycle. But then it was found that each of these Dansgaard–Oeschger cycles was but one part of a larger pattern of climate change. That larger view came from a different kind of evidence coming from cores not of ice but of sediment and rock.

The pioneer in these studies was geologist Gerard Bond of the Lamont-Doherty Earth Observatory, who compared ice-core records with sediment cores obtained by deep-sea drilling. He reasoned that the changes in climate recognized from the ice cores should show up in the sediment cores, with different kinds of sediments found in the times of cold compared with the times of warmth.

To gather his data, he laboriously counted the number and kinds of benthic foraminifera found in sediment, which leave behind their microscopic shells after their death. Other researchers had already found that certain species of these forams, as they are more familiarly called, were found on the bottom of the coldest ocean water, while others were typical of warm water. Bond thus had a crude but useful paleothermometer. While this method could not tell exactly what the temperature was where the shells were formed, another method, using oxygen isotope data, could. But just looking at the percentage of warm versus cold species was much faster than the laborious and expensive analysis of thousands of individual forams found at different levels, and hence different times of the core. On land and in the ocean, Bond observed that after a particularly large warming event—the next three, four, or even five Dansgaard–Oeschger cycles—showed a progressively dropping mean temperature in similar areas of the cold part of the cycles. With each cycle the subsequent rapid warmth was less than the same event of the preceding cycle.

Finally, the coldest of the following cycles left other evidence of an even colder, non–Dansgaard–Oeschger interval. In these coldest of cycles, there were many small pebbles and cobbles left on the seafloor, and these were named Heinrich layers in honor of German researcher

Hartmut Heinrich, the geologist who first observed them (but at the time could not explain this curious phenomenon). While such material does on occasion reach the deeper sea through underwater landslides caused by turbidity currents or grain flows, the very large number of the large broken rocks, or clasts, indicated that icebergs had passed overhead, melted, and dropped the rocky load they had acquired as glaciers growing downward through rocky areas.

The sediment cores showed layers that were made up of almost nothing but ice-rafted pebble layers. Periodically the extent of glacial ice on the continents and ice caps was so large that armadas of icebergs filled the Atlantic. Good thing that no *Titanic*-like ocean liners were around in those times, for the North Atlantic would have been spectacularly filled with the floating icebergs. Bond then noticed that the rapid warming following one of these very cold ice-rafting events caused global temperatures to shoot up to values higher than those in any of the preceding warm parts of the Dansgaard–Oeschger cycles. These would have been the highest relative global temperature swings, abrupt climate change to warm conditions of staggering magnitude, and they are now called Heinrich events.

The longest view was not available. For the last 100,000 years, 90 percent of the time the climate was very cold, allowing the growth of vast glaciers on the northern continents. The warm episodes between longer periods of cold were quite short, centuries perhaps. But a really curious aspect was noticed as well. Very anomalously, the cycles over the last 10,000 years were few and of low magnitude change. Our planet has been in a strangely long-lasting, 10,000-year interval of warmth. Yet six times in that last 100,000 years, there were Heinrich events; six times the great iceberg armadas filled the North Atlantic, cooling the air around them, bringing about the coldest times, followed rapidly by the warmest times.

Thus by the middle part of the last decade of the twentieth century, the massive amount of data leading to the recognition of cycles was joined with models of the conveyer belt. Was there a relationship between parts of the Dansgaard–Oeschger cycles, and even the time of iceberg armadas, with the flickering on and off of the North Atlantic conveyer currents? The answer from many different models by various research groups was a resounding yes. The conveyer belt seemed to shut down during the cold parts of the cycles, and then start up and stay running during the warm intervals. Over the last 10,000 years, most of them warm (there have been some minor warming and cooling periods, as we will see in Chapter 8, "The Oncoming Extinction of Winter,"), the conveyer current in the Atlantic was running continuously, and if there were shutdowns, they were of short duration. But there is a chicken-and-egg problem here. Did the changing conveyer alter climate instead of being a victim of climate change brought about by some other factor? That question has yet to be answered.

What of our second question: Is the conveyer changing in some way today? The answer to that very important question is still unknown, but early data suggest a very scary yes. For the first time in our time, a research group has reported what it claims is a slowing of the most important of the Atlantic currents, probably due to massive amounts of fresh water entering the sea in northern areas because of the rapid melting of the northern ice cap. This report, published in 2006, is the first overt link to massive volumes of fresh water coming from melting Arctic ice and its effect on the conveyer.

Many scientists, including Richard Alley, in his now classic and important 2000 book, *The Two-Mile Time Machine*, regard the Atlantic conveyer current system as very finely balanced and hence very susceptible to change. The easiest way to cause this change, according to sophisticated computer models, is to pump in fresh water into the northern part of the system. The truly staggering—and just now realized—melting of Arctic ice, a story not even noticed prior to about 2003, is pumping in massive volumes of fresh water at the most dangerous place for the integrity of the conveyer.

We may be seeing the start of a changeover that has now been recognized as having happened repeatedly up to 8,000 years ago, and then stopped. The conveyer system in its present configuration has thus been stable for a significant amount of the time that humans have had agriculture, and this stability has allowed both predictability of crop yields in Europe and Asia, as well as the biologically more important stability of ecosystems. Ecologists have long known that organismal diversity rises with stability. It is rapid change that leads to loss of biomass as well as biodiversity, with the end-point being mass extinction itself.

Perhaps it is the on and off of the conveyer belt that tips Earth's climate one way or the other. If new volumes of fresh water suddenly enter the system, causing the conveyer belt to turn off, it causes the Earth to rapidly cool. If the conveyer belt is turned back on, sudden warming happens. This is the pattern that seems to have occurred during the ice ages, when there is significant ice at both poles. But perhaps there is a second possibility, one that builds the bridge linking past with present. What if the conveyer belt does not turn off but stays on and changes the place where the cold, fresher surface water sinks to the depth from high in the north to more southerly midlatitude regions? What if we are looking at two entirely different kinds of conveyer belts (and their workings)—one typical of the Ice Age we are in, and one from older, warmer, ice-free times? This latter kind is what seems to have happened in the Paleocene epoch, and as we have seen, the consequence was that deep, cold, and oxygenated bottom water from high latitude sank and changed to deep, warm, anoxic water. The aftereffects were mass extinction. If, through some act of God (or act of humans), the Earth warmed to the point that the ice disappeared entirely, then a new set of models have to be used. It would literally be a whole new (old) world.

The key to climate change seems to be both the level and the rate at which carbon dioxide rises in the atmosphere. It is that information that finally bridges us to the present day.

CHAPTER 8

The Oncoming Extinction of Winter

MAUNA LOA, HAWAII, 2001

A sparkling ocean showed the blue-green of the shallow reefs lining the tropical shoreline. The chopper headed inland, passing over lush green forests as it aimed toward the still distant volcano. This was a route rarely taken by any of the numerous tourist helicopters buzzing routinely above the active and recently inactive volcanoes of Hawaii. Mauna Loa, located on the "Big Island," is now extinct, having not erupted in thousands of years. But people still flock to it in swarms of buzzing helicopters, beginning their flying tours of it from some higher point than even the nearest coast, for the cost of gas makes a journey from sea to peak prohibitive. For most, anyway. This trip was on the dime of the medium with the deepest of all pockets: not tourism, or science, but television.

Soon the helicopter and its crew had reached the mountain's vast base, and vegetation dropped away, revealing raw, naked lava slopes with extravagant, toothpaste-squeezed deposits of lava from the last eruptions, a jagged rockscape difficult for any human to scramble over. Now the slope of the volcano steepened, altitude was quickly gained, and a curious new kind of deposit was now beneath, cobbles and dropstones from some ancient glacier, now long since gone.

As the helicopter reached the 10,000-foot level, both humans and machine were affected by the altitude. The humans, all acclimated to surface-level oxygen values, found themselves gasping a bit, breathless already after speaking to one another while excitedly pointing out feature after feature of the novel flight. The helicopter was also feeling the lowered atmospheric pressure. Its rotors chewed the thin air ever more futilely, and the straight-line passage from the coast to the top of the mountain changed to a series of long, slow switchbacks necessary to keep climbing. Finally, at 12,000 feet, the research station came into view, and good thing, for the helicopter with its greater-than-normal load of humans and cameras was like a gasping athlete staggering to the finish line.

Gratefully dropping onto the landing pad, the crew looked at the nondescript buildings making up the station. It had been built in the late 1960s with only one mission: to monitor the amount of carbon dioxide in a place where local human efforts would not cause anomalous readings. This place was one of many such stations scattered around the globe, but it had pride of place as the flagship of the bunch.

The reason for this expensive ride was documentary television, in this case Discovery Channel Canada, with an experienced, traditional filmmaker named Tom Radford, making an old-style documentary about mass extinctions and the future of Earth's biodiversity, and like the subject, the show was to be rich in content, with long "talking head" segments, no MTV-style, machine-gun editing, no *Jurassic Park*–like monster animation. Radford had booked this copter, bringing his cameraman and soundman, and I was the "talent," or talking head, for a standup in front of one of the highest, least-known, yet important research stations on the planet, the carbon dioxide measurement lab located nearly atop the gigantic Mauna Loa Volcano on the Big Island of Hawaii.

The crew and I offloaded and began laboriously carrying gear toward one of the huts. The whole place seemed deserted, a white pickup at the end of the twisting road from below, now parked against one of the buildings, the only evidence that anyone was at home. A lonely job this was, to take these measures at constant intervals; most of the work was now automated, but lighthouses still needed human keepers; a staff was still needed to make sure that the crucial readings of atmospheric carbon dioxide levels were dutifully recorded.

Laboring in the high altitude with bags, tripods, and cameras that at sea level were so easily carried, we found the one staff member then on duty. He seemed quite used to seeing a film crew, for as the last years of the twentieth century came to a close, more and more film crews made the trek up to this lonely outpost to see firsthand what carbon dioxide levels were doing year by year. By the end of the twentieth century, there was a four-decade record. The stationmaster showed the film crew the equipment but then told them that there was one place in the station that all the other filmmakers eventually gravitated to: On one wall was a simple graph, showing the raw data measured by the station. It was simple because the graph showed a single trend—upward. But superimposed on that line rising toward the 400 parts per million levels that the future would inevitably bring in surely only a few short years from its then current level of about 350 parts per million was a more subtle relationship. The slope of the line steepened, approaching the present day. Not only were carbon dioxide levels obviously rising but they were also rising ever faster as time went on. The implication of that was not lost on the filmmakers as they interviewed their somewhat breathless talent, who tried to explain the graph in talking head fashion. A simple graph, with perhaps terrifying ramifications (Figure 8.1).

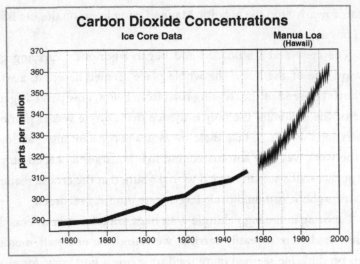

FIGURE 8.1

Change in carbon dioxide levels since beginning of the Industrial Revolution

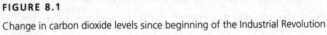

That the amount of carbon dioxide in our atmosphere, as evidenced at the carbon dioxide laboratory, has increased is undisputed. What is disputed, however, is what has caused this rise, and what (if anything) it will mean to global climate.

Few, except those who for political or economic reasons (such as representatives of the big oil companies and the politicians that they have bought off), dispute that humanity is rapidly changing the composition of the atmosphere (although there is still great debate about whether those changes are causing a rise in mean global temperature, also known as global warming). The carbon dioxide is largely coming from automobiles and human industry. These anthropogenic, or human-induced, sources of gas go beyond carbon dioxide: There are also methane, chlorofluorocarbons, sulfur dioxide, and nitrogen oxides, the levels of which have been rising dramatically since the Industrial Revolution. All of these gases have the ability to absorb infrared radiation and reradiate it back to Earth, producing the well-known greenhouse effect. Predictions about the possibility of future global warming over the next decades and centuries to come from a class of models known as global circulation models (GCMs). A starting point of these models is the prediction that the amount of carbon dioxide in the atmosphere will double over the next century. Most climate scientists agree that this doubling is sure to have profound ecological effects, including greater temperature increases in midlatitude, temperate, and continental interior regions relative to the rest of the globe; decreases in precipitation in these same midlatitude regions; and an increase in severe storm patterns.

The debate about global warming can be categorized as follows: first, that it is not happening; second, that carbon dioxide and other greenhouse gases are rising but that this rise is caused by volcanic sources, not people; third, even though greenhouse gases are rising, they will have no effect on current and future climate. Thus the naysayers. A 2004 novel written by the inimitable (and curiously science-hating) mega-author Michael Crichton, *State of Fear*, uses each of the points above to argue against any sort of human-caused global warming. But what do the climate scientists, not the authors and politicians, say?

A new computer model developed by scientists from the University of East Anglia in England has factored in the role of human-made global warming. This model suggests that the human input of greenhouse gases will indeed delay the next ice advance by perhaps as much as 50,000 years—but that when it does arrive, it will be an even more extreme and longer period of ice than otherwise might have occurred. The amount of hydrocarbons that can be burned is finite, and sooner or later they will run out.

Although many experts do think that human-produced global warming could postpone the next ice advances by many millennia, there is another school of thought suggesting that the rapid global warming that is now underway may actually *trigger* the next glacial advance. According to this model, seemingly paradoxically, the advent of global warming now could kick us back into the time of ice. Let us look in more detail at the carbon dioxide curve, and another curve as well—the methane curve—and let us take a less obvious perspective—that anthropogenic global warming is not just a two-century phenomenon but one that began soon after humans discovered agriculture.

TEN THOUSAND YEARS AGO, HUMANS BEGAN TO SHIFT FROM HUNTER gathering to farming, from isolated bands to living in cities by the new crops, from small population numbers to larger and better-fed populations. Doing so brought with it a new kind of atmospheric input: human-created methane.

This proposition is the major thesis of veteran climatologist William Ruddiman in his 2005 book *Plows, Plagues, and Petroleum*. Ruddiman readily conceded the obvious: that the last two centuries have witnessed an unusual rise in carbon dioxide and methane levels. But Ruddiman takes a longer view that the more recent rises are sitting on the shoulders of more ancient human events, noting, "Slower but steadily accumulating changes had been under way for thousands of years, and the total effect of these earlier changes nearly matched the explosive industrial era increases of the last century or two." This can be analogized to the great volcano in Hawaii where so much of our information about carbon dioxide rise has come from: Mauna Loa rises 14,000 feet above sea level, and while it is an imposing mountain, at this maximum elevation it is about half the elevation of the Everest-like peaks. But the subaerial, visible portions of Mauna Loa are but part of the mountain. In reality more than half its height is covered by the sea; taken in sum, Mauna Loa is far higher than Mount Everest. So too with the greenhouse gas record—recent rise is like the subaerial parts of Mauna Loa, but the rises of the past 10,000 years, covered by the seas of time, are as much a part of the story as the more visible recent rises.

Ruddiman breaks climate history into three phases. The first extends from the earliest times on Earth to about 8,000 years ago. Over this staggeringly long time, as Ruddiman brusquely puts it, nature was in control. Then, at the mark of 8,000 years before the present, give or take some centuries, for the first time good old Mother Nature had some competition for creating atmospheric change. (Of course, other organisms have been involved in atmospheric change. But we humans are the first to do it with technology, rather than our own physiology.) The final phase, beginning about two centuries ago, marks an acceleration of trends from the 8,000-year mark onward. Thus, in this description of climate through time, Ruddiman maintains that humans have been changing the atmosphere far longer than is generally supposed.

What is the support for this interesting view? The details come from one of those interesting mergers of scientific fields that occasionally occur (too rarely, actually) but that often result in important insights derived from the edges of scientific fields. In this case it is the merging of anthropology and climatology that has led climatologists such as Ruddiman and Brian Fagan, in his delightful 2003 book *The Long Summer*, to think outside the box, as well as lending power to Jared Diamond's *Guns, Germs, and Steel* and his recent *Collapse*. The subtitle of Fagan's book is *How Climate Changed Civilization*, and that is the theme of Diamond's books as well. Ruddiman, however, would say that the subtitle should be *How Civilization Changed Climate*—and as we all now know, civilization continues to change climate.

At the heart of this new history is the timeline of human agriculture. The earliest evidence discovered to date comes from the Fertile Crescent region of Mesopotamia, and in the Yellow River region of

China. Agriculture began in other regions after this, but in some cases it was thousands of years later, such as in the Central American lowlands, the Peruvian Andes, and the tropics of Africa. The two regions first out of the gate benefited from both a climate amenable to cultivating plants, as well as the presence of an abundance of edible plants that themselves could be domesticated. In the Fertile Crescent these included wild barley and rye, peas, lentils, and other grains and cereals. There were also mammals with biological traits that made them amenable to domestication, such as wild goats, sheep, pigs, and cattle. At the same time there were raw materials necessary to produce tools and implements required for agriculture, such as appropriate material to make scythes, mortars and pestles, and baskets for carrying the crops from harvest to village.

And there was far more going on than just raising the first crops. For the first time, humans began deforestation on a planetary scale, through either burning the forests or logging.

Even with the first rudimentary agriculture, it took another 1,500 years—up to 10,500 years ago—for the first permanent villages to appear, and there is soon after the first evidence of animal domestication, but from there things progressed quickly, both in the increasing sophistication of agriculture and its rapid spread around the world.

And with that spread, Ruddiman argued, humans began to affect the atmospheric gas concentration. He proposed (first in a scientific paper published in 2001, later expanded in his 2005 book) that by 5,000 years ago, an anomalous rise of the potent greenhouse gas methane began, and it has continued ever since. The chief source of this methane may have been from the flooding of vast lowland regions to allow the cultivation of rice or the diversion of rivers for other primitive irrigation attempts. In these new wetlands, large volumes of plant material decayed, died, fell into low-oxygen settings, and were, through decomposition, converted into many reduced organic compounds, including methane. Livestock also produced methane on scales that exceeded its production from natural sources, which is mostly either volcanic in origin or the emission of methane gas from gas hydrate (frozen methane) sources.

The third phase, ushered in with the Industrial Revolution, was two-pronged. The burning of vast quantities of coal sent great volumes of carbon dioxide into the atmosphere. At the same time, the cutting down of forests increased in tempo.

Even if we accept Ruddiman's evidence (and others' as well) of an earlier than acknowledged role in the rise of atmospheric carbon dioxide, this rise is seemingly inconsequential compared to the wholesale emission of carbon dioxide and methane as a consequence of the Industrial Revolution, coal economy, and automobile oil economy. But was it really? Here Ruddiman presented a bombshell of an idea: that had humans not begun agriculture, there would now be a gigantic, continental ice sheet covering regions of northeastern Canada. Because climate change occurs with feedbacks—as it gets colder (and so there is more ice), the albedo, or reflectivity, of Earth increases, causing it to become colder yet. By now the Northern Hemisphere would be well into the cooling cycle that builds continental glaciers—a new ice age—if not for agriculture.

So what do the numbers say? Using laboratories such as that featured at the start of this chapter, we humans can very accurately keep track of carbon dioxide in the atmosphere, and we can even measure the amounts in ancient atmospheres by minute sampling of trapped gas bubbles in the ice cores. We saw in Chapter 6 (Figure 6.1), "Bridging Deep Past and Near Past," that the carbon dioxide record since the evolution of animals some 550 million years ago was one of both rises and falls. In general, the amounts of carbon dioxide in the past have been much higher than that of today.

But let us look at this record not in terms of millions of years ago, but in thousands. The Vostok ice core from Antarctica has yielded a very detailed record of carbon dioxide levels over that time. It shows that carbon dioxide levels varied between a minimum of 180 parts per million and a maximum of 280 parts per million. Thus, for more than 200,000 years (and actually going back nearly 2 million years, in fact), atmospheric carbon dioxide values (and methane values as well, which mirror those of carbon dioxide) seesawed up and down, and as they did, global temperature went up and down as well. If we break down carbon dioxide levels into either above or below an arbitrarily picked level of about 240 parts per million, it turns out that the lower half of the cycling carbon dioxide levels in aggregate were occupied for more time than did the higher half. During the low–carbon dioxide times, the Earth accumulated the great ice sheets—we were in the Ice Age.

Breaking out of the range of 180 to 280 parts per million did not happen until about 1800, when carbon dioxide levels began to rise well beyond the old upper limit. By 1900, the level was 295 parts per million, or an increase of about 15 parts per million in a century. But that was just the warm-up, so to speak. From 1900 to 2000, carbon dioxide levels went from that 295 parts per million all the way up to about the current level of 370 parts per million—a rise of *80 parts per million* in the last century—and the curve described by these data gets ever steeper. The rise will continue as China and India join Europe and the Americas in putting two cars in every garage and heating many new houses with natural gas and oil. Even if we stayed at a rise of 80 parts per million over the next century, by the year 3000 the atmosphere would have a carbon dioxide level of about 450 parts per million. But most atmospheric scientists use the rate of rise over the last 50 years, rather than the last 100, to predict the future amount of carbon dioxide. Using those rates, which work out to about 120 parts per million per century, we might expect carbon dioxide levels to hit 500 to 600 parts per million by the year 2100. That would be the same carbon dioxide levels that were most recently present sometime in the past 40 million years—or more relevant, *it would be equivalent to times when there was little or no ice even at the poles.*

Yet climatologists seeing the newest data emerging are now dismissing even that scenario. Carbon dioxide increase into the atmosphere is accelerating. Models using the latest values of rise for the past decade, and projecting forward, lead to an estimate that carbon dioxide levels will quadruple. And it will not take a millennium to do it. By 2200 we might expect to see carbon dioxide levels approaching 1,200 parts per million. In as little as a century levels will be approaching 1,000 parts per million.

Greenhouse gases strongly affect planetary temperature. As carbon dioxide levels rise, so will planetary temperature. Because the heat budget of the Earth is complicated by the effects of the oceans, land, and especially currents (water and air), there is not a linear relationship between carbon dioxide rise and global temperature. The rule of thumb used by climatologists is that each doubling of the carbon dioxide level can be expected to increase global temperatures by about 2 degrees Celsius. Thus the projected carbon dioxide level even for a century from now would be expected to increase the global temperature between 3 degrees and 4 degrees Celsius. Today that temperature is estimated to be between 15 degrees and 16 degrees Celsius. It would climb to just beneath 20 degrees Celsius. The effect of that would be Earth-changing, conceivably bringing about the greatest mass death of humans in all of history.

What about the lethality of these gases themselves? The activity of these gases directly kill by carbon dioxide or methane toxicity, or by producing a by-product effect of their rising levels in the atmosphere: global heating. To this we can add another potentially lethal by-product: acidification, a process we have not yet addressed here. To do this we have to digress briefly into ocean chemistry.

Carbon dioxide reacts with various other molecules in many kinds of reactions. Several of these are directly involved in maintaining the acid or base levels of the ocean. Two chemical species, the bicarbonate ion (HCO_3)$^-$ and carbon dioxide, form part of the chemical buffer system that maintains a relatively neutral level of the oceans (neither acid nor base). However, if atmospheric carbon dioxide rises, the ocean becomes more acidic through a chemical reaction leading to formation of hydrogen ions in the sea (that is what acid is—lots of hydrogen floating around in solution). We measure the concentration of this hydrogen ion level using the so-called pH scale, with lower values corresponding to higher acid levels. A rise in acidity, at small levels, poses no danger to organisms. But if the levels rise enough, organisms are directly threatened. Rising acidity is most dangerous to organisms that produce calcareous shells, such as coral reefs and a type of phytoplankton called coccolithophorids. Also, once the acid levels rise, they stick around at high levels for a long time: The ocean pH change will persist for thousands of years. Because the fossil fuel–induced rise in carbon dioxide is faster than natural carbon dioxide increases in the past, the ocean will be acidified to a much greater extent than has occurred naturally in at least the past 800,000 years.

We are sure that over most of geological time, the carbon dioxide level in the atmosphere was higher than it is now. Does this mean that the oceans were more acidic than now? At least for the last 100 million years, this was probably not the case. If there is lots of calcium carbonate in the upper reaches of the ocean (as there is when there are abundant blooms of the organisms that make chalk, such as the coccolithophorids and the foraminifera), that can strip the carbon dioxide, and so the excess acid, out of the water. But the buffering takes time, and that is the biggest difference between the rise in carbon dioxide and its effect now compared with anytime in the past. During slow natural changes, the carbon system in the oceans has time to interact with sediments and so stays approximately in steady state with them. For example, if the deep oceans start to become more acidic, some carbonate will be dissolved from sediments. This process tends to buffer the chemistry of the seawater so that pH changes are lessened. But what humans are doing in terms of injecting carbon dioxide into the oceans from emissions is unprecedented. *The present rise in carbon dioxide levels seems to eclipse any other rate of increase from the past.* It is this rapid increase that outstrips the natural buffering systems, resulting in oceanic acidification. Thus it is unlikely that the past atmospheric concentrations would have led to a significantly lower pH in the oceans. The fastest natural changes that we are sure about are those occurring at the ends of the recent ice ages, when carbon dioxide levels increased about 80 parts per million in the space of 6,000 years. *That rate is about one-hundredth that of the changes currently occurring.*

Our world is hurtling toward carbon dioxide levels not seen since the Eocene epoch of 60 million years ago, which, importantly enough, occurred right after a greenhouse extinction.

CHAPTER 9

Back to the Eocene

The town of Bellingham, Washington, is one of those smaller American cities that routinely makes the various "most livable" lists. And why not? It has a good state university, ensuring an influx of culture and scholarly lectures unavailable in most cities of its modest size. It also sits amid stunning green hills rising out of the cold, clear waters of Puget Sound. The gigantic volcano Mount Baker looms over it in regal fashion, and the rainy but cool climate ensures year-round verdure. Most of its trees are scrubby, deciduous maples and alder, with the numerous garden transplants such as rhododendrons, camellias, and magnolias adding spring color. But as recently as a century ago, the vegetation had a radically different look. In place of the now dominant flowering, broad-leafed trees there rose gigantic Douglas fir and western red cedar in old-growth splendor. Needles, not leaves, reached skyward for photons, and so dense was this forest that its floor was in perpetual gloom, to the depressive chagrin of the still-settling inhabitants. Soon enough they felled these trees and saw the sky.

This grand forest was a bristly blanket that stretched several thousand miles along the Pacific Coast, gradually changing species composition northward toward Alaska, and south into the Northern California coast. And it is ancient, not just in the short measures of human history but also in the more robust and virtually unimaginable scale of millions of years. While buffeted by the ice ages, advancing with the warmer intervals only to again retreat back into small pockets during the height of the ice sheets, the western North American coastal forest stretches far back into the nebulous geological past. It is old—but it sits on a rock cover that is older yet, holding evidence of a very different West Coast than we know now.

Chuckanut Drive is a beautiful, windy road leading southward from Bellingham along a steep rocky coast. It is often closed by seasonal rock falls, for the road was carved into steeply dipping sedimentary strata, dating back some 60 million years, a time when vast mountains to the immediate east were rising upward, and in so doing rapidly eroding and dumping vast volumes of gravel, sand, and mud into the rivers, streams, and accompanying ponds, swamps, and lakes in the forelands. The grains of these sedimentary rocks give away the provenance of their origin, telling of mountains made of granite and high-grade metamorphic rock. But more interesting than these clues to their rocky origin are other clues to a past 60 million years old. Most of the precipitous outcrops along Chuckanut Drive bear bedding planes smeared with fossil leaves. In some places the rock seems to be nothing but bedded leaves, as if some ancient gardener willed an autumn leaf compost pile into rock.

Broad-leaf fossils in a place now that should harbor nothing but pine needles? And not just any broad-leaf species—the fossils show the exotic and exuberant leaf shapes that today are found only in the humid tropical jungle of our world. And if any doubt remained about the heated time that these Chuckanut formation fossils come from, they are immediately erased at every larger bedding plane, for imprinted over the smaller tropical leaves are gigantic, spectacular palm fronds. Washington State, meet Florida.

Vertebrate paleontologists have also prowled these beds, finding numerous turtle shells and crocodile fossils. And just as the Douglas fir old-growth forests should today extend vast distances along the coast, so too did this ancient, 60-million-year-old forest from the time interval formally known as the Eocene epoch cloak huge areas of North America. The Eocene palms and crocodile fossils can be found as far north as the Arctic Circle. There is only one possible explanation for this distribution. In the Eocene epoch, the world had to be far warmer than it is today. Warm enough to allow tropical flora and fauna to thrive in what is now the land of permafrost and ice.

What would it be like to live in such a world? What if all of human civilization was suddenly transported to the Eocene world? Our coastal cities would be in for a nasty, wet surprise, for they would be instantly drowned. The Eocene epoch was so hot that there were no polar ice caps, and thus sea level was about 150 feet higher than it is today. There would be no snow angels or autumn leaves or sledding for any American. There would be no seasons at all, other than endless summer. Today, even Los Angeles and Miami detect some passing of the seasons. Not so in this Eocene world. Where should we go now to see such a world?

A long jet-plane hour east of Australia lays the island of New Caledonia. Beautiful place, this Calédonie (as the locals call it). It is a huge island somewhat parallel to, if south of, the Great Barrier Reef off Australia. It is not your run-of-the-mill Pacific atoll chain, typified by a low topography made up of crushed white limestone gripping a coconut palm community in equatorial heat, but a real hunk of continental land, high mountains being the backbone of 300 miles of 75-mile-wide real estate, one of the biggest islands in the world, with two really different and sensational kinds of rocks raising it from ocean depths. There is lots of limestone, of course, for the entire length of the island is skirted by a wide and fabulous barrier reef of diverse Indo-Pacific coral, while the 10-mile-wide lagoon made by the outer barrier reef is a veritable carbonate factory. But that is just veneer, for this island is both old and something peculiar on Earth's surface: It was torn from the ancient, Permian–Triassic supercontinent of 250 million years ago, ripped from its Gondwanaland heritage by the titanic tectonic forces that created the Atlantic Ocean and at the same time sent all the continents scurrying to new places about the globe. New Caledonia was but one small sliver, but in the tearing, it scooped out rocks from far deeper in Earth than is the norm, rocks from Earth's mantle region itself, that deep place on which the peripatetic crustal plates float. The rocks of this region are far denser than their silicate-rich cousins on the surface, with a far higher metal content. New Caledonia became a slice of metal ore, eroding to deepest red in color when eroded to soil. It was rapidly colonized by European powers once its mineral wealth became known, and it is still a colony of France to this day, one of the last. It is featured here because it gives us a glimpse of what the future world may look like. Even though, as we saw in Chapter 6, "The Driver of Extinction," our oncoming carbon dioxide levels are more akin to those of the Cretaceous period than the Eocene epoch, the similarity in flora and fauna of this latter time interval makes for a more accurate comparison.

At first glance, that future seems like a pretty good deal, especially for those who live in colder climes. New Caledonia is not on the equator—far from it. It straddles the Tropic of Capricorn, latitude 25 degrees south, and because of this its water is cooler than many places of lower latitude, yet still warm enough to support coral reefs and many varieties of palm trees. It has a huge barrier reef that encloses many smaller reefs gathered in the wide lagoon, and around and within these smaller reefs lives a rich molluscan fauna. Among many varieties of the more prosaic bivalves there is a high diversity of snails spectacular in their extravagant color and morphology. Cones, conchs, turrets, whelks, turbans, and more—even the rare chambered nautilus—can be found off the deeper reefs. Yet even among these many beauties there is one family of snails that stands out in terms of pleasing color and shape: cowries, those colorful snails living only in warm water.

Thus with its palms and snails, New Caledonia can serve as a vision of what much of the globe might be like, at least for a geologically short time in the future as our planet warms, and its animals and plants are already familiar to those paleontologists and paleobotanists studying the 60-million-year-old Eocene epoch. Their fossils are common in the numerous and rich Eocene-age deposits found in many places around the world, including those where even during the Eocene the animals of the time were living at high latitude, places quite cold in our world but warm enough for the New Caledonia kinds of animals in the past.

The Eocene epoch, a time of warmth. It is this ancient Eocene that many of the experts looking at and projecting Earth's future climate now study, for the Eocene was the last time the world was totally globally warmed to worldwide tropical conditions with palms and crocodiles and cone shells and nautiloid cephalopods spanning the globe, a time when there was absolutely no ice at the poles and snow was something limited to the highest mountains only, a world where once again palm trees, tropical mollusks, and basking crocodilians will be able to make a living in places like Canada or northern Europe, as the world once again becomes a tropical paradise.

Will we be going forward into the past? In this final chapter let us look at what the Eocene epoch was like, in order to prepare us for what a world with carbone dioxide levels of 1,000 parts per million will be like.

THERE IS A TRUISM NOT KNOWN TO MANY WHO LIVE THEIR LIVES IN the temperate realms of the world. Even those in places such as Florida think that they live in heat, and that they do during the summer months. But the occasional frost still menaces the Florida citrus crops; there are many cool and comfortable days in winter. And besides, the summer of Florida and every other hot but industrialized place of human habitation keeps the heat at bay with the vast air-conditioning enterprises that eat up so much of our planet's energy output each year. These places do not qualify as really hot places. The *really* hot places are united by a very different human activity than turning up the air conditioner: Because human life is so miserable in humid, unrelenting equatorial heat, *everyone* uses drugs, drugs to help escape the heat, the misery, to make time go by. We who live in the more comfortable climes seem to think that just because the human tribes who have long inhabited the equatorial zone have evolved through many generations living in constant heat, night and day, that somehow these people no longer feel the heat and humidity, that unlike us, they are not made uncomfortable by the horrible climate. Not so. Hence the many varieties of "little helpers" to get through the day.

They are not called drugs, of course, nor are they illegal. But an interesting variety of pharmacological substances can be traced around the world in the world's hot zones, and the habit of using these various drugs goes back through history in each place that they are found. Starting in Fiji, and heading east through the Western Pacific Islands, the drug of choice is kava root. In every market stall or place of work outside or in, the Fijians invariably have a coconut or other kind of bowl with the milky white liquid within. Drinking goes on all day, and the effect is to make time pass more quickly. To get through the day, in other words.

Moving east to Vanuatu, the old New Hebrides Islands, the kava gets immeasurably stronger. Instead of the Fijian variety that provides a pleasant buzz to the point that one forgets the heat, the Vanuatuans are nearly knocked over by their potent brew, which has an awful taste, but one sure isn't bothered by the temperature of 98 degrees Fahrenheit and the 99 percent humidity. In Micronesia, the drug changes. Here, and up into the old Indochina peninsula, the drug of choice is betel nut. This nasty stuff also yields a potent buzz. It is ingested by chewing the tough little nuts wrapped in a small bit of palm leaf with some white coral grit enclosed as well. The calcium carbonate reacts with the alkenes in the nut to form a red intoxicant that is not swallowed but swished around the mouth and then spit out in a highly staining red expectorant. Sidewalks, roads, market stalls—all are stained red because of the habit, as are the gums of the chewer. A more significant change occurs as well: The prolonged chewing of the coral grit grinds down the teeth, often to sharp points, and a smile from a red-mouthed, sharp-fanged Micronesian, spitting out gobs of red goo, puts to shame the special effects of any Hollywood vampire. The effect of the betel nut buzz is heightened by smoking island marijuana, a shrub now found growing wild throughout the widespread islands of the many archipelagos stretching from the Philippines through the vast regions of Indonesia.

As one moves into India, the drug of choice again changes, to khat, an intoxicant widely available. As in the other tropical areas, its use is something that goes on all day and cuts through most social classes. Again, it provides a pleasant buzz, and while the heat of the day remains, the day is conquered and the unpleasantness of the heat is put aside. Khat is widespread in Asia Minor and is found as well in the Sahara regions. The more vegetated parts of Africa at equatorial latitudes have khat but many other kinds of drugs as well, as befits a place where the diversity of plants and people is so high.

Finally, swinging around the world again, now to South America, we find the widespread chewing of coca leaves as a way to beat the heat. Hours on end, spitting out the used leaves and chewing new ones, the day goes by unnoticed, energy levels are increased in the enervating heat, the day's duties are accomplished.

We humans evolved near the equator, it seems, but this brain of ours does not do well when heated for long periods of time. British neurobiologist Martin Wells, the grandson of H.G., once observed that human thinking is best done at "sweater" temperatures—in other words, if it is cool enough to require a sweater, as is the case in older British households, characterized by a lack of central heating. There may be something to this; the sum total of great intellectual insights and contributions coming from the extreme tropics is scant indeed, as are the products of prodigious human industry in such places. No brand of cars comes from any country with year-round heat; no computers, no airplanes. Southern India, which has long hot seasons interspersed with monsoons, is becoming the sole exception to this, but its factory work and thinking take place in air-conditioning, not outside in the heat. Singapore is also an exception, an equatorial powerhouse that exists as such only because of the most extensive use of air-conditioning to be found on the planet, at huge cost in energy.

There is another characteristic of the equatorial regions: malaria. While AIDS remains the most visible killer in the hot zones, far more people die of malaria, and the infection rate throughout most of the really tropical areas is staggering. For example, in the Solomon Islands (a very hot island chain in the western Pacific), the infection rate is more than 90 percent. The *Anopheles* mosquito is the vector of this protozoan-caused malady, and fortunately for humanity this mosquito requires great heat to live. One can only imagine what humanity would be like if the many species of mosquitoes in temperate and Arctic regions also carried and caused malaria. Heroic efforts are underway to reduce the misery of malaria, but all efforts at a vaccine have so far failed, and the current prophylactic measures involve ingesting poisons toxic enough to kill the protozoa in the human bloodstream but not quite toxic enough to kill the human. This is a very poor solution, and sooner or later most visitors to the tropics will contract this killer.

TWO QUESTIONS NOW ARISE: WILL THERE BE ANOTHER GREENHOUSE extinction similar in any way to the events of the deep past profiled in this book? If one is in our future, when might it occur? For the moment let us accept an affirmative answer for the first and see what (if any) consensus there already is regarding the second.

The latter question was examined in a landmark paper published in *Nature* in 2005. That study estimated that climate changes brought about by global warming will lead to the extinction of more than a million species by the year 2050. Since there are only 1.6 million species now identified (although many more are yet to be described), such numbers result in an extinction rate of more than 60 percent. To compare this with the past, this number would place the next greenhouse extinction second only to the Permian extinction. And the first million species, if the *Nature* study is correct, would just be the start of things. As we shall see below, a shift to a new kind of oceanic conveyer current system would create an anoxic ocean, eventually changing into a Canfield ocean. The shift from mixed to anoxic ocean would likely kill off the majority of marine species, just as it has in each of the ancient greenhouse extinctions.

With this in mind, let us return to the first question posed above. Can such an event be already happening—are we in the first stages of a greenhouse extinction? For this latter question, our knowledge coming from the past extinctions is of little use. The rock record is excellent at tracking million-year or even hundred-thousand-year events. But here we are looking at events happening on decadal scales. There is no ice-core equivalent in the rock record that resolves such short-term events in the past. Yet we can gain insight into this question by looking at the state of the world's climate in the present.

Books take time to write and time to be put into print. Any book is a multiyear effort. In 2006, as I write words that will not appear in print until 2007, we can try to summarize the state of Earth's climate in 2006. (By *state* we might mean the picture produced by the values of temperature and greenhouse gases and the nature of the conveyer belt, among many others.) Hopefully the state of the climate will be about the same in 2007 as it was in early 2006. There will be more carbon dioxide and methane in the atmosphere, of course, and more of the ice caps will have melted, freshening the sea, most dangerously in the North Atlantic. But perhaps the rate of change is faster than one can hope, fast enough, perhaps, to have taken our world past the combined climate tipping point. Of all of the irreversible changes that might be triggered by the tipping point, two are paramount—the oceanic conveyer, obviously, given the importance it has been awarded throughout this book, but also the great ice sheets now resting atop Antarctic bedrock or Arctic land and sea. Ice sheets on Greenland and Antarctica hold 20 percent of all of the fresh water on our planet, water locked up in its solid phase. But what happens if all of that ice melts?

Let us look at this and additional environmental changes that could lead to the next greenhouse extinction, including sea-level rise, ocean acidification, global warming (oceanic as well as terrestrial), and coral reef "bleaching."

THERE WAS PLENTY OF HEATED CONTROVERSY IN 2006 ABOUT WHETHER the high-latitude ice bodies are already on an irreversible slide toward melting. It turns out that the early phases of the irreversible slide will be masked by natural process and because of this, proceed very slowly. With a warming atmosphere, the edges of ice sheets melt and glaciers recede. But the melting does not all go into the ocean. Local climate change and the warming itself can increase rainfall over the ice caps, and if this precipitation reaches the cold central regions of any ice body or begins to fall anytime in the still-frigid high-latitude winters, it falls as snow, which rapidly is converted back into the freshwater ice that this water originally came from. The edges melt, the center accumulates new ice, and the system only slowly moves toward something much more dramatic. At some point the warming ocean, the source of all this change, increases in temperature enough to cause disintegration of the ice sheets. Faster and faster they melt, first calving off armadas of icebergs and later simply converting to water, which finds its way into the ocean. The ocean freshens, but more ominously, the volume of new liquid water entering the world ocean is so great that the very level of the oceans themselves, known as sea level, begins to rise.

The rise in sea level that has occurred to date is still very low, on the order of a centimeter over the last century. But if either part of the Antarctic (western part) or all of the Greenland ice sheet melts, which would occur (according to climate models) with a global rise in temperature of between 2 degrees and 3 degrees Celsius, the rise in sea level would be 6 meters, or about 20 feet! If both melt, the rise is more than 60 meters, or 200 feet. Good-bye, all coast cities, and good-bye, a good proportion of the planetary agricultural yield, since a very significant quantity of human food is grown in the large deltas such as those found at the ends of the rivers Nile, Mississippi, and Ganges. All of the deltas and their rich soil would be pretty well inundated with even a 1- to 2-meter rise in sea level. The eventual rise of 25 meters would bring back the old coastlines of the Eocene epoch.

Melting of the ice sheets would produce a radically different climate than what we have now. Radically different. As stressed here, what we call climate is made of many individual and largely interconnected systems, and the past evidence of change suggests that these thresholds are both sensitive and can have dramatic consequences, once a critical level is passed. A good way to analogize this is by thinking about the action of an electric light switch. Slowly increasing pressure on the button does nothing until the threshold is reached, and once that point arrives the switch jumps forcefully and quickly into a new position. Pushed past a threshold, most climate systems can jump quickly from one stable operating mode to a completely different one.

A rising sea level would be the most dramatic effect of ice-cap melting. But in all probability, no less important would be the consequence of all of that freshwater entering the oceanic conveyer belt system. As we saw in Chapter 5, "A New Paradigm for Mass Extinction," the conveyer is powered by the density and temperature difference of its seawater at different geographic areas and depths. Freshwater entering the system in the North Atlantic would be particularly significant. South of Greenland is the area where previously warm Atlantic Ocean seawater, which had made its way from the tropics off the Caribbean, finally cools enough to sink into deep water. Warm water has more salt ions, and once it cools, its density is higher than surrounding water. But the injection of fresh water, with a much lower density because of its lack of salt ions, would effectively stop the conveyer or perhaps shift where it starts and stops on the surface. A rising sea level would drown cities, but a conveyer belt shift would kill people, lots of them, because of the great effect it would necessarily have on climate in European agricultural areas. It can be surmised that a suddenly cooled, cropless European subcontinent with its large population would by necessity look toward still-arable lands to make up food loss. Here's hoping under this scenario that the Europeans have enough cash in reserve to buy an awfully large volume of food for centuries to come.

The rise in sea level displaces not only crops but people as well. This is an aspect so obvious that it is usually lost in any discussion of the effects of rising sea level. However, as any urban geographer can attest, a large proportion of humanity currently resides in coastal or low-elevation riverside locales. All such localities would be affected by even a small rise in sea level, and when we start looking at 25-foot increases (a common estimate for an ice-free world following melting of the Greenland and Antarctic ice sheets), we see a reality in which vast populations of humans will have to move to higher ground. Perhaps nowhere is this more evident than in the low-lying country of Bangladesh, which currently has one of the densest populations of humans on Earth and whose population is estimated to double over the next century. Let us look in detail at what a 25-foot rise in sea level would do to that country.

While it seems at first glance easy to map a future coastline following a known rise in sea level, simply by making the new coast at the appropriate topographic level on a detailed map of the region, in reality such mapping is more complex than that. Coastal areas are prone to subsidence—sinking as the wet soil beneath them compacts—while the flooding of deltas, lagoons, estuaries, and especially the river mouths of large continental rivers can produce startlingly different topography. One group that has attempted to make maps taking these factors into account is the future-mapping group at the University of

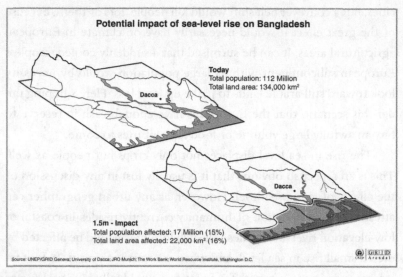

FIGURE 9.1

The impact of a 1.5-meter rise in sea level on Bangladesh

Arizona. The mapper in chief, geographer T. Overbeck, has put online a number of such maps, and these are reproduced here with his kind permission.

In Figure 9.1, Overbeck shows the current geography and population centers of Bangladesh. Currently, Bangladesh is home to 112,000,000 people on 134,000 square acres of land. What happens to these people and the land area with a rise in sea level? The bottom part of Figure 9.1 shows the estimated new shoreline positions after a rise in sea level. In this case, however, the map is based not on the catastrophic maximum rise in sea level of 25 feet but on *only a 5-foot* (1.5-meter) rise, which all scientists agree is inevitable, largely because of the expansion of the oceans from their warming but also because of the initial volume increase from the ice that has melted to date. That rise in sea level would displace 17 million people (15 percent of the population) and inundate 22,000 square acres—16 percent of all land area.

So what happens with a rise in sea level of more than 25 feet? Because Bangladesh is so low-lying, this kind of rise would almost wipe out the entire country. Only a small strip abutting the Indian subcontinent would remain subaerial. Virtually the entire population of Bangladesh, one of the poorest countries in the world, would have to migrate. But who would take the perhaps 200 million people who would need land, food, water, and energy on an unprecedented scale?

The Bangladesh case brings home the urgency of confronting this problem. With global help, countries like Bangladesh could probably cope with the 1.5-meter rise in sea level. And there are plenty of other countries in similar straits. Indonesia, for instance, has great areas of its habitable land surface of such low elevation that it would be largely flooded by the higher of these two rises in sea level.

Let us look at another case—the United States. Again, we can map the areas that are within 1.5 meters of sea level, as shown in Figure 9.2.

From the areas in black, it is clear that large parts of Louisiana, Florida, and estuaries along the Atlantic coast, especially Delaware Bay, would be covered by a 1.5-meter rise, and far more by a 3-meter rise. South Florida, the population center of the region, would be especially hard hit. The higher rise in sea level, not shown in Figure 9.2, would of course prove far more catastrophic.

All in all, it is safe to say that between one-quarter and one-half of all people on Earth would be displaced by the 25-foot rise in sea level. Words cannot begin to suggest the human suffering and mass extinction of humans that would occur. Our world cannot let the ice caps melt. But have we already passed the tipping point, at least for Greenland?

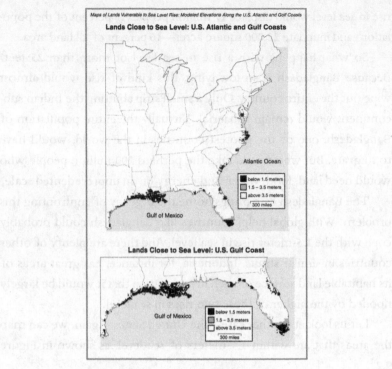

Maps of Lands Vulnerable to Sea Level Rise: Modeled Elevations Along the U.S. Atlantic and Golf Coasts

Lands Close to Sea Level: U.S. Atlantic and Gulf Coasts

Atlantic Ocean

below 1.5 meters
1.5 – 3.5 meters
above 3.5 meters

300 miles

Gulf of Mexico

Lands Close to Sea Level: U.S. Gulf Coast

Gulf of Mexico

below 1.5 meters
1.5 – 3.5 meters
above 3.5 meters

300 miles

FIGURE 9.2

The effect of a rise in sea level on the United States

Let us look at new and ominous data on glacial movement in Greenland that point toward a more rapid reduction in ice cover than previously considered.

WHILE RELATIVELY SMALL COMPARED WITH THE AMOUNT AND THICKNESS of ice found on the Antarctic continent, the Northern Hemisphere ice caps, and especially the ice cover on the subcontinent of Greenland, hold a formidable volume of water as ice. Since the ice-cap ice floats, its melting has no effect on sea level. Not so for Greenland, however, where the ice sits on rock, not seawater. As we have seen, if all of the Greenland ice cap were to melt, the sea level would rise 6 to 7 meters, or about 20 feet. Because Greenland is closer to the equator than Antarctica is, the temperatures there are higher, so the ice is more likely to melt. And not only is air temperature higher around Greenland than above the Antarctic continent but the temperature of seawater around Greenland is also higher than that of seawater around Antarctica. The crucial observation that needs to be made is whether the ice on Greenland is melting, and if it is, how fast.

This is where the alarming new data come in. In early 2006 a study determined the average rate of movement of the glaciers on Greenland (most ice there is tied up in glaciers, which are slow-moving rivers of ice). While melting of an ice cap conjures up pictures of an ice cube disappearing on a Phoenix street corner in July, the reality is that melting also involves the rate at which the glaciers, most terminating at the coastline, dump ice into the sea. Much of the now water-borne ice floats off as icebergs. The study showed that the speed of the glaciers had increased by a factor of eight compared with a decade earlier. This more rapid speed could only be caused by lubrication at the base of the ice—and this lubrication is water, whose source is indeed the melting of ice in the traditional manner. If the glaciers can replace the ice they lose to the sea at the same rate, there is no net loss. But the opposite is happening. The glaciers are not being replaced at a higher rate. In fact, some are not being replaced at all. Every indication is that Greenland is poised to see its ice cover disappear with increasing speed. The north Arctic region has undergone a regional temperature increase that is 20 times that of the whole world. That is what is driving the disappearance of the Northern Hemisphere, and especially Northern Hemisphere, high-latitude ice—such as that of Greenland.

How long till all is gone? Not in the twenty-first century, perhaps, even at the rapid rate just measured. But enough will disappear to certainly have an affect on sea level. Perhaps the 2-foot rise predicted is woefully underestimated. And as we have already seen, even that slight rise will negatively affect crops and people, especially in the highly productive deltas of the world.

Antarctica holds the world's main ice cover, with about 90 percent of the world's ice (and 70 percent of its fresh water). Antarctica is covered with ice an average of 2,133 meters thick. So what would happen if we lost all ice caps? If all of the Antarctic ice melted, sea levels around the world would rise about 61 meters—about 200 feet. That is where I think we will be by the year 3000.

THERE IS ALREADY SIGNIFICANT HUMAN MORTALITY FROM THE CURRENT greenhouse-induced global warming of Earth. A 2004 study by scientists at the World Health Organization and the London School of Hygiene and Tropical Medicine determined that 160,000 people die every year from the effects of global warming, from malaria to malnutrition, children in developing nations seemingly the most vulnerable. These numbers could almost double by the year 2020.

A second cause of human mortality comes from storm-related deaths. Any suggestion that better technology for forecasting could reduce the danger of oncoming storms through earlier evacuations was certainly exposed as myth by the tragedy of Hurricane Katrina and its flooding of New Orleans as well as vast tracts of coastal Louisiana by oceanic storm surge. As Earth's tropical regions become warmer, its systems of redistributing that heat become more energetic. Thus, the warmer this planet gets, the warmer the Atlantic Ocean gets, bringing warmer and more moist ocean air, the fuel of hurricanes. This is why scientists and insurers fear climate change will worsen hurricanes. The number of the deadliest hurricanes—that is, category 4 and 5 hurricanes—has, between 1990 and 2004, almost doubled since the period of 1970 through 1985. Globally, between 1990 and 2004, there has been an increase from an annual average of 10 such hurricanes to an annual average of 18. The increase in intensity of hurricanes is the direct result of an increase in water temperature of 0.5 degree to 1 degree Fahrenheit. While some argue that a natural, 30-year cycle in hurricane number may be part of this cause, it is also true that this 30-year periodicity may itself have been affected not just by the last few decades of global warming but also in fact by two centuries of rising carbon dioxide levels.

Another danger to human life is heat waves. Heat waves in August 2003 caused 35,000 deaths in Europe, 15,000 of them in France alone. The U.S. Environmental Protection Agency points to one study that projects in New York City the probability of warming of 1 degree Fahrenheit, which could more than double heat-related deaths during a typical summer, from about 300 in 2006 to more than 700. The lead author of this study, Thomas Karl, director of the National Climatic Data Center, noted in his summary of the situation:

> It now seems probable that warming will accompany changes in regional weather. For example, longer and more intense heat waves—a likely consequence of an increase in either the mean temperature or in the variability of daily temperatures—would result in public health threats and even unprecedented levels of mortality. High temperatures are likely to become more extreme, and because night temperatures will increase by at least as much as daytime temperatures, heat waves should become more serious.

Already we have seen killer heat waves that caused more than 500 heat-related deaths in Chicago in 1995 and more than 250 deaths in the eastern United States during a period of hot weather in the fall of 1999. And things are projected to get worse. The World Meteo-rological Organization projects that by the year 2020 there could be 3,000 to 4,000 deaths in the United States alone. These numbers will be dwarfed by human mortality in cities in other nations that are less energy rich. As fuel costs of cooling increase, the number of poor dying globally because of heating periods will skyrocket.

CLIMATE CHANGE IS ALREADY THREATENING THE PLANET WITH THE spread of infectious diseases, which will move farther northward and to higher elevations. The World Health Organization projects tens of millions more cases of malaria and other infectious diseases than exist now. While insects are proliferating, carrying these diseases, three-fourths of all bird species are on the decline. We are thus losing our first line of defense against the threat of disease-carrying insects, since insectivorous birds are the major insect predators. In addition, 26 percent of bat species are threatened with extinction. It is estimated that bat colonies in Texas alone eat 250 tons of insects each night. The loss of many species of birds and bats, while insects proliferate, could lead to an escalation of the use of pesticides, threatening yet more damage to the world's animal species, including us.

The misery of living in a tropical climate as well as the ever-present threat of contracting malaria are the two aspects of climate change through heating that don't get much press. Yet as the tropics begin to spread north and south from the low latitudes of Earth, scourges of the tropics will be coming too. We are returning to a planet with worldwide malaria foremost, but there're more: Ebola, elephantiasis, schistosomiasis, leprosy, rampant intestinal parasites, poisonous spiders and centipedes, new and vicious kinds of ants—all will follow the heat once the barriers of coolness are overcome.

WHILE THE PROBLEMS FOR HUMANS LISTED ABOVE ARE SERIOUS ENOUGH, they are not the two most lethal dangers. The greatest threats posed by global warming are surely famine and war, two Horsemen of the Apocalypse going hand in hand.

Our world sits on a knife edge of global starvation already. We six billion humans, heading toward a far higher number at about the time that rising carbon dioxide levels should begin to stabilize a new pattern of climate, are able to be fed, all of us right now, through the miracle of that long-ago breakthrough of the human mind, agriculture. We need every bushel of grain, however. There cannot be even a single season without harvest in either hemisphere, and this is why there is extreme danger of rapid weather change if there is a Krakatoa-type volcanic explosion or impact of a 100-meter or larger asteroid. Both would put so much dust in the air that one hemisphere or the other (or perhaps both) would have a yearlong or longer winter and thus no crops.

Short-term climate change would be nearly as devastating, and in the long run, more devastating. Neurobiologist Bill Calvin, who has written extensively on the dangers and effects of sudden climate change, suggests that a 10- to 20-year event is far more difficult to deal with societally than is a sudden catastrophe.

Why would a warmed world be in danger of plummeting crop yields? It would seem that plants might flourish in the higher carbon dioxide levels, and with longer growing seasons, perhaps an additional crop could be counted in many areas. This will surely be true for some kinds of human food. Tropical fruits and starches will be available in abundance. But the staple of human sustenance, grains and cereals, the very first crops, in fact, from 10,000 years ago, would suffer. The grain belts rely on cool but not frigid winters, and summers with abundant moisture. Current projections are that the great breadbaskets of Earth, especially the greatest of them all, the American Midwest, would have climate changes that would reduce summer moisture. As droughts become more frequent, yields of wheat, corn, barley, and oat crops would decline.

In the new climate, new regions would become arable that currently are not. Two thousand years ago, northern Africa was the granary for the Roman Empire, but climate change since then caused an expansion of the Sahara Desert and dryness in the formerly fecund states of Morocco, Tunisia, Algeria, and Libya. Those regions would likely get more rain and could perhaps again begin producing bountiful harvests. But it is not likely that they could immediately take advantage of the more propitious climate. Efficient farming is highly mechanized and highly oil intensive. All of the African states listed above are Muslim countries with some of the highest population growth rates on the planet. They do not have a tradition of American-style megafarms, the institutions that create the current food surplus that are so important to help feed so much of the world. They do not have factories that can manufacture the complicated farm machinery necessary. The same goes for areas in Eastern Europe, and all of sub-Saharan Africa. South America could pick up some of the slack, but not all, should the American Midwest become a dust bowl of greater extent than during the Great Depression of the 1930s.

The second great problem is warfare. Nations are unlikely to sit around and watch their populations starve or their national treasuries deplete in order to buy enough food. It will become more and more tempting to simply take or blackmail other countries with nuclear weapons. The desert kingdoms and dictatorships of the Middle East, watching their deserts become even more arid, will become increasingly dangerous as many become armed with nuclear weapons.

The next two centuries will be an interesting time. Our ingenuity as a species could let us get through this. Our darker natures and impulses, however, in the face of sudden climate change, could result in the loss of half of all humans on Earth in a century or less.

AT WHAT LEVEL WILL GREENHOUSE GAS LEVELS PLATEAU AND THEN DEscend, and, more important, how much will the world warm? Our homework, then, is to ensure that the world warms no more than 2 degrees Celsius from its present state. Why is that goal important, and how realistic is it? A guest column by Malte Meinshausen, Reto Knutti, and Dave Frame in the best source for climate change—realclimate. org—on January 31, 2006, is a good discussion and summary of this problem (and offers a possible solution), and if I pirate the spirit of their article, it is (hopefully) for a good cause.

The three authors go through the math, showing that a stable carbon dioxide level of 400 parts per million (to reiterate, we are at about 380 parts per million and rising as I write this in 2006) will yield an 80 percent chance that Earth will warm no more than 2 degrees Celsius. For instance, the rise from carbon dioxide levels of 280 parts per million at the start of the Industrial Revolution to the present level of 380 parts per million has brought about a global temperature increase of 0.8 degrees Celsius, thus calibrating the climate models used to predict future temperature increases that are tied to greenhouse gas concentration increases. The good news is that one of the most troublesome of greenhouse gases now being produced by human activity, methane, has a short life in the atmosphere before it breaks down. Also, the oceans are an effective sink for atmospheric carbon. If human emissions can be sharply curtailed in the twenty-first century, concentrations of all greenhouse gases could begin to decline near the end of the century according to the best models now available. However, these are just that—models. This model even lets greenhouse gas levels peak to 475 parts per million for a short time, but we do not go past the 2-degree increase if we can then bring them back down to 400 parts per million before the end of the century.

So how does society do this? Drive less. Drive less-polluting cars. Buy hybrids or electric cars. And there is more. For instance, the authors state:

> We need to start taking large amounts of carbon out of the air. One very good way to get this going with positive environmental effects if managed properly is to grow biomass then char it and use the elemental carbon to mix in large quantities deep into soil as "terra preta" or Amazonian dark earths. This has major soil conditioning properties (i.e., reducing conventional fertilizer need by 50 percent).

In other words, use the enhancing carbon dioxide levels to grow lots of new plant material, turn that biomass into charcoal, and bury it into tropical soils.

WITHOUT HOPE THERE WILL BE NO ACTION. AS FAR AS CAN BE SEEN IN the present, we have not yet reached the point of no return, or the tipping point. We as a worldwide society can keep carbon dioxide levels below 450 parts per million. If we do not, we head irrevocably toward an ice-free world, which will lead to a change of the thermohaline conveyer belt currents, will lead to a new greenhouse extinction. The past tells us that this is so.

FINALE

The New Old World

We sat high in a smallish office, a cubicle like that of most academics, littered with books, articles, files, individual sheets of paper. Like the shallow subtidal marine world of the tropics, it was a place where every square inch of surface was covered, not by animal life as in the ocean but by ground-up plant life turned to pulp, paper, ink, and knowledge. The man I talked to was tall, spare, disheveled in an academic way; colleagues of his kept butting in asking for graduate students' files, and I tried to keep up as David Battisti spun images of a new old world.

It was a lecture by this man, back at the end of the twentieth century, that stimulated this book. In that lecture, given not to students but to other science professors at the University of Washington, Battisti said that current climate models were inadequate to explain how the 60-million-year-old Eocene epoch of the deep past was so warm with the carbon dioxide levels that had been found to occur back then, and that we were heading for those same Eocene-like levels in the twenty-first century. I had come to ask him if he still held those startlingly radical views, and as he talked and I scribbled, a whole new view of things became clear to me. Battisti's work has been featured earlier in this book, and he is one of the modern architects of climate science. He is a fitting guide to end this book and for seeing how the end of our familiar world will play out as well.

Yes, we are still heading for Eocene-like carbon dioxide levels. As shown in Figure 6.1 in Chapter 6, "The Driver of Extinction," the Eocene epoch had carbon dioxide levels of about 800 parts per million. And yes, he reiterated these many years later, we will hit that level by the end of the twenty-first century. I replied back with the hope with which I ended the last chapter, the societal hope, that we can hold the line at 450 parts per million. Battisti laughed out loud at that, a mirthless laugh at the inanity of that hope, a forlorn hope, for Battisti, like me, and surely like so many of you readers, has children. How about in the century after that? I asked. He frowned, mused, showed his infectious grin. It would be 1,100 parts per million, he said, because the destruction wreaked by 800 parts per million will finally have caused society to do something. But even that something, the real curb of emissions, will slow, not stop, the rise in carbon dioxide and other greenhouse gases into our world's atmosphere.

What about the ice sheets? I asked. After writing this book, I have concluded that the world's ice sheets are going, going, to be gone, leaving us with an ice-free world. Quite right, he said. Greenland first. Then Antarctica. How long for Greenland? I asked. I give it about 300 years at most, he replied. And Antarctica? Longer, was the reply, but it too will go to an ice-free condition, probably by the end of the millennium, for there is a pile of ice down there. But it will go. And then we really will be back to the Eocene.

Time was ticking by; I knew this man was frightfully busy. He had received the offer of a professorship at Harvard, the ultimate compliment, but eschewed that offer, deciding to stay in the Pacific Northwest and part of a team of people at the University of Washington at the forefront of climate research, and much good, societally and scientifically, was the result. I felt embarrassed to be taking so much of his time, but he was not squirming or looking at the clock; he began talking faster and faster, carrying us into the new old world, and in the process alternately fascinating and scaring me. The inadequacy of the models to explain why a world with carbon dioxide levels of 800 parts per million could have been warm enough to allow crocs and palms in the Arctic and Antarctic was brought up. So what is wrong with the models? I asked. Clouds, he replied. The models do a very poor job of simulating clouds. Clouds are the wild cards, controlling opacity of the atmosphere to light, changing albedo, Earth's reflectivity, but also, if in the right (or for society, in the wrong) place, they act as super greenhouse agents. It is in very high parts of the atmosphere, the altitude where jumbo jets cross the world, where the change in clouds will be most important. Global warming could produce a new kind of cloud layer, clouds where they are not currently present, thin, high clouds, higher than any found today, completely covering the high latitudes and affecting the more tropical latitudes as well, but even that is a misnomer, as most of Earth will have become tropical at that time.

Take me there, I said, and he did, a verbal journey. We started first in the Arctic, in winter.

Trees can now grow everywhere, but all their leaves are gone, because we are in the months-long winter night. There is some light coming from a filtered full moon. There are no low clouds to be seen, but the moon is almost obscured by hazy high clouds, and the moonlight has an unfamiliar cast to it. There are no stars, and Battisti tells me that the haze above is high and ever present. There would be no starry nights, and, in summer, no perfectly clear days. High haze and high, thin clouds would see to that. A most notable aspect of this Arctic world is presence of lightning, gigantic bolts that seem to come from nowhere, and as my eyes adjust to the dim light, I can see that many of the trees are blackened, from fire. The surface is warm, but in the long night the air aloft is cold even in this globally warmed world, and lightning is common.

He then took me south, to the midlatitudes where most of the world's population lives now, in our time, to Seattle, in fact.

The city that I had been so familiar with is gone: The Space Needle is now a 400- rather than 600-foot monument, emerging from the sea like a societal middle finger directed at the human generations before that had created this world. Here too the sky is different, but this is daytime, and its color has changed. The distribution of plants and the omnipresence of dust in the summertime due to the drying of the continents in the midlatitudes has changed the very color of the atmosphere; it is strangely murky as yellow particles merge with the blue sky to create a washed green tinge, a vomitous color, in fact. Gone too are the periodic cold and wet fronts that hit the Pacific Northwest every three or four days in winter. These storms are gone, the climate tranquil, kites a thing of the past in this world. Palm trees are everywhere.

Finally, on to the tropics, and here there is nothing but destruction. Unlike the midlatitudes, where storms have subsided to a calm tranquility, here the violence of hurricanes has only increased. The tropics have warmed, which breeds more ferocious storms, but storms with shorter tracks, no longer menacing the regions that had once feared them most. They stay confined in the tropics, but because the world has warmed, there is far less wind shear now, and wind shear is a tamer of hurricanes. But with it gone, they are unchecked, monster storms, category 5 hurricanes now the norm, and newer, higher categories have been invented. There are no crops here, and there is little human habitation.

To the North Atlantic, to see the conveyer system. In the twenty-first century, it had stopped, for some decades, and Europe had indeed cooled. The alarmists had predicted quite wrongly that Earth was finally sliding into the much overdue ice part of the ice ages, but this cooling was regional to Europe and short; the rocketing carbon dioxide levels saw to that, shooting well past any chance of a global glaciation as so often happened over the last two million years. But the conveyer had not stayed shut down for long; now it was chugging away, but in a far different geography than before. The superheated warm water of the tropics headed north as before, but the sinking happened well south of its original Greenland location. Now the vast quantities of water slightly cool while heading north and sink in the mid-latitudes, and the water sinking is very different from the cold, oxygenated water of before. This is warm water that no longer can sink to the abyss, and thus the delivery of oxygen to the bottom of the ocean has stopped. The deep ocean is now a graveyard, warm deadly water bathing species that had evolved and adapted for something quite different, and the start of the mass extinction is already under way. The ocean is returning, rapidly, to its most common ancient state, the anoxic state, and already poison is accumulating on the bottom, hydrogen sulfide concentrating, year by year.

How did we get to this future? I asked. Easy, Battisti said.

In the late 1960s and early 1970s, the amount of carbon dioxide in the atmosphere increased by 1 part per million per year, just like clockwork. But every year from then on, the rate increased, until by the year 2000 it was increasing by 2 parts per million per year. By the middle part of the twenty-first century, it was increasing at 4 parts per million. The reason was simple. The vast multitudes in India and China had all demanded, and bought, a car for every house, and were now moving toward two cars in every garage, as their North American and European fellow world citizens had long enjoyed. Now two cars in garages were appearing in most houses in the Middle East and North African shores, places with the highest birth rates on the planet. In the twenty-first century, the human population hit nine billion, and a goodly percentage of them drove to work each day.

An hour had gone by. I was back among the stacks of paper with Battisti. I had one last question. I need to close the book, I implored, and as yet my last chapter just seems to end flatly. He mused, then asked if I knew the stages of acceptance that anyone diagnosed with a fatal disease goes through: first denial, then anger, then action, and, if that action fails, the final acceptance before the final event itself. I did not see the connection. Look, he said; think about the major environmental problems faced and ultimately solved during the twentieth century. The ubiquitous presence of DDT, for instance. Rachel Carson, in her masterful book *The Silent Spring*, most famously alerted the world to the dangers of this chemical. Change ultimately occurred; different pesticides were ultimately used. So too with many of the victories in the United States; the Clean Air Act did clean the air to a chemistry far more healthy for humans in large cities. The Clean Water Act did help reduce toxins in the water. The Endangered Species Act did save species. In each case there were defeats, but in each case, victories were won, year by year.

I was still not getting it. What did this have to do with the current problem of global warming? I asked. Battisti was quick: Each of those environmental victories already had a political system or structure in place that could implement the required changes, at least in the United States. And because the United States was the main producer and exporter of so many environmental toxins, the changing of rules there resulted in improvement globally. But that is the main difference with the global warming threat from those other examples, Battisti explained. At the present, there is not a political system in place that can—realistically—accommodate and accomplish the necessary changes. What is necessary, he said, is a true global system for implementing regulations and economic incentives that will, on a worldwide basis, lead to emission reduction. That is a pipe dream now, but as the world warms and climate rapidly changes, that too will surely change.

My time was up. I had a notebook crammed with new facts. I bid him good-bye. That was fun, he said, having this discussion. Best time I have had this week. And then a look flickered across his face, the realization of what that new old world that we had constructed in our talk would look like, do, affect, change. Fun to talk about? Yes, but then came guilt with the realization of what the "fun" translated into.

WITH MY INTERVIEW WITH BATTISTI FINISHED AND AFTER COMPLETING the changes to the manuscript for this book asked for by Battisti and geochemist Eric Steig, both of whom so gracefully read the manuscript in search of bonehead errors, I considered myself through.

Weeks passed; spring began to come to Seattle with ever warmer days, and not for the first time it seemed that the flowering trees and emerging buds were coming forth ever earlier in the calendar year. It seemed that what was needed to end this book was a number of possible scenarios about the future. Here they are, inelegant but variably plausible, based on all that has come before in this book and the massive scientific literature dealing with global warming and climate change.

Scenario 1: This is the status quo—the most hopeful and perhaps the best that we can hope for, a scenario in which humankind does reduce carbon emissions sufficiently to keep the atmospheric carbon dioxide level below the target threshold of 450 parts per million, the somewhat arbitrary figure mentioned in Chapter 9, "Back to the Eocene," as the level that we should not exceed.

At first glance this scenario seems plausible enough, looking solely at the rate at which the level of carbon dioxide is increasing in our atmosphere. Whereas at the start of the 1900s the rate of increase seems to have been about 1 part per million per year, the construction of modern carbon dioxide–sensing labs, such as the one that we saw in Chapter 8, "The Oncoming Extinction of Winter," tells us that the increase is now about 2 parts per million per year. As I write this, that level is about 380 parts per million. With 94 years left in the century and assuming that the rate of rise would be the same 2 parts per million per year, a simple calculation puts the level of carbon dioxide at the start of the twenty-second century at 548 parts per million. This is well above the goal of 450 parts per million but well below the 800 to 1,000 parts per million that David Battisti's models see coming by that same date, the year 2100. To ensure that we do not pass the level of 450 parts per million, then, which is only 70 parts per million higher than today's values, our society would have to go back to the increase of 1 part per million that was ours before cars. Let us assume that we somehow manage to do this. What would the world be like at the start of 2100? Again, wishing and hoping that there are not the upper-atmosphere complexities seen by Battisti and others, we would have a world where sea level would have risen "only" about 1 meter. The conveyer belt current system will not have stopped. The ocean stays mixed. A greenhouse extinction has been averted.

Is this a pipe dream, with the rapid and currently ongoing industrialization of China and India, as well as of other populous countries previously lumped in the "Third World" category? In terms of climate change, the Third World countries have little industrialization and few personal cars. But the number of such places is diminishing. Human population is increasing, and so too are global standards of living. Countries such as Indonesia, Mexico, Burma, Thailand, and even Vietnam are rapidly producing a middle class. All will want to drive.

Scenario 2: Here let us assume that the countries above join the club currently inhabited by the United States, Japan, and Western Europe—the club of prodigious greenhouse gas emissions. Let us say that carbon dioxide levels hit 700 parts per million by the year 2100. What will the world look like?

Global temperatures will have risen by 2 to 3 degrees Celsius, or perhaps by 5 or more degrees Celsius. The greatest change is to the northern ice cap. A tenth of the ice previously on Greenland is gone, and the great ice packs of the North Pole region are now open sea, to the delight of shipping lines. But this is as nice as the sea has become. Globally it has risen by 2 meters—more than 6 feet. Millions of people in Bangladesh are just now being displaced into an ever-tighter corridor in the highest elevation of their low country. Other countries with sea-level elevations are facing the same problem. The sea has begun most importantly to encroach on the port regions of major cities. And more ominously, the North Atlantic conveyer, in the space of a decade, has shut down. The ocean bottoms are quietly beginning to accumulate reserves of reduced carbon, and the first deep, benthic species of foraminifera are decreasing in number in response to the anoxic bottom water as they inexorably head toward extinction. But the world has taken little notice of forams as the last wild populations of polar bears disappear. Europe is experiencing frigid winters and has lost many of its important cash crops, including its entire wine industry (and just when everyone could use a good drink!). Globally wheat production is down and entire forests are changing due to the die-off of many species intolerant of the sudden changes in their regions, some from too much cold, some from too much heat, and many too from the change in annual rainfall amounts and patterns. The first stages of a warm monsoon are becoming evident in the Pacific Northwest, while the Great Basin and southeast are becoming ever more parched deserts.

So scenario 2 looks pretty bad. Yet that is but a harbinger of what could happen and will eventually happen even under that scenario if the increase in carbon dioxide levels is left unchecked. Like the Ghost of Christmas Future, let us look at what may be a worst-case scenario.

Scenario 3: The arbitrary year of 2100 is a time when many pent-up feedbacks and checks on climate change have now been overwhelmed, and 2100 is but a station between two very different worlds. The level of carbon dioxide is at 1,100 parts per million. Global temperatures have risen by more than 10 degrees Celsius. Greenland is half exposed, and the great Antarctic ice sheets have begun their own melting. The world is now racing toward an ice-free world. All of the world's seaports are drowned, their former populations driven inward. In hindsight, we know that the disastrous flooding caused by Hurricane Katrina, that now long-ago storm, signaled the advent of the new hothouse world now just coming into its malignant first flower. It was just a foretaste of what was to come when all U.S. ports would flood, as they have now, when Galveston Island is entirely underwater, as is most of South Florida. And still the water rises, and it will continue to do so as long as Antarctica keeps melting, its complete disappearance predicted to occur about 900 years in the future. There is no central government in the United States in anything but name. The states have reverted to tiny nation-states, hoarding and grappling with the immigrants streaming in from other states, especially in the Midwest, where food can still be grown and fresh water is still available.

The long-predicted shutdown of the North Atlantic conveyer system happened, but it did so rapidly, and then started up again—but in new fashion. As at the end of the Paleocene epoch, its start was in equatorial regions, but the downwelling was in newly warmed midlatitude regions, and the water flooding downward onto the ocean basins contained far less oxygen than surface water from previous times in this place. In response, the deep ocean has rapidly come to resemble the bottom of the Black Sea. The first licks of anoxia are rising to the surface in some regions as well, as scientists measure and worry about the first appearance of the hydrogen sulfide–producing bacteria.

Chaos is global. Tunisia was fought over, its ancient Roman granary regions again producing some of the highest yields of wheat in the world. The combatants were the Franco-German alliance. Half of the world population was forced to live on the minimal wheat-equivalent diet necessary for sustaining human life. Half of the population was also on the move, and those moves meant war. World War III was fought over high land, food, and water, and the strongest grabbed. Tel Aviv, Tehran, and Marseilles were radioactive craters from nuclear attacks bent on settling old scores. Luckily for Earth, no one now knew how to make an atomic bomb anymore, and the many old ones still around were rapidly becoming unusable. But not rapidly enough for some millions of humans.

Chaos reigns, but if humans suffer, at least their species is not endangered. Not so for so many animals and plants. Ten percent of all species on Earth are now extinct, after yet another greenhouse extinction.

Specific References Alluded to in Text

INTRODUCTION: GOING TO NEVADA

Guex, J., A. Bartolini, V. Atudorei, and D. Taylor (2004). High-resolution ammonite and carbon isotope stratigraphy across the Triassic-Jurassic boundary at New York Canyon (Nevada). *Earth and Planetary Science Letters* 225(1–2): 29–41.

Ward, P.D., J.W. Haggart, E.S. Carter, D. Wilbur, H.W. Tipper, and T. Evans (2001). Sudden productivity collapse associated with the Triassic-Jurassic boundary mass extinction. *Science* 292(5519):1148–1151.

CHAPTER 1: WELCOME TO THE REVOLUTION!

Alvarez, L.W., W. Alvarez, F. Asaro, and H.V. Michel (1980). Extraterrestrial cause for the cretaceous-tertiary extinction—experimental results and theoretical interpretation. *Science* 208(4448):1095–1108.

Benton, M.J. (1993). Reptilia. In M.J. Benton, *The Fossil Record*. London: Chapman and Hall. 2:681–715.

Bohor B.F., P.J. Modreski, and E.E. Foord (1987). Shocked quartz in the Cretaceous-Tertiary boundary clays. *Science* 236:705–709.

Hallam, A., and P.B. Wignall (1997). *Mass Extinctions and Their Aftermath*. Oxford: Oxford University Press.

Keller, G. (2003). Biotic effects of volcanism and impacts. *Earth and Planetary Science Letters* 215:249– 264.

Keller, G., W. Stinnesbeck, T. Adatte, D. Stueben (2003). Multiple impacts across the Cretaceous-Tertiary boundary. *Earth Science Reviews* 62:327–363.

Kring, D.A., and D.D. Durda (2002). Trajectories and distribution of material ejected from the Chicxulub impact crater: Implications for postimpact wildfires. *Journal of Geophysical Research* 107(6):1–22.

Kyte, F.T., J. Smit, and J.T. Wasson (1985). Siderophile interelement variations in the Cretaceous-Tertiary boundary sediments from Caravaca, Spain. *Earth and Planetary Science Letters* 73:183–195.

Margolis, S., J. Mount, E. Doehne, W. Showers, and P. Ward (1987). The Cretaceous-Tertiary boundary carbon and oxygen isotope: The stratigraphy, diagenesis, and paleoceanography at Zumaya, Spain. *Paleoceanography* 2:361–377.

MacLeod, K., and P. Ward (1990). Extinction pattern of *Inoceramus* based on shell fragment biostratigraphy. *Geological Society of America Bulletin* 247(special paper):509–518.

Marshall, C., and P. Ward (1996). Macrofossil extinction patterns in K/T boundary sites. *Science* 274:1360–1363.

Melosh, H.J., N.M. Schneider, K.J. Zahnle, and Latham D (1990). Ignition of global wildfires at the Cretaceous/Tertiary boundary. *Nature* 6255(343):251–254.

Melosh, H.J., and A.M. Vickery (1991). Melt droplet formation in energetic impact events. *Nature* 350:494–497.

Mukhopadhyay, S., K.A. Farley, and A. Montanari (2001). A short duration of the Cretaceous-Tertiary boundary event: evidence from extraterrestrial helium-3. *Science* 291:1952–1955.

Muller, R. (2002). Measurement of the lunar impact record for the past 3.5 b.y. and implications for the Nemesis theory. In C. Koeberl and K.G. MacLeod, eds., Catastrophic Events and Mass Extinctions: Impacts and Beyond. Boulder, Colorado: *Geological Society of America* (special paper) 356:659–665.

Phillips, John. *Life on the Earth: Its Origin and Succession*. Cambridge and London: MacMillan, 1860.

Raup, D. (1991), *Extinction: Bad Genes or Bad Luck?* New York: W.W. Norton.

Raup, D.M. (1992). Large-body impact and extinction in the Phanerozoic. *Paleobiology* 18(1):80–88.

Raup, D.M., and J.J. Sepkoski (1982). Mass extinctions in the marine fossil record. *Science* 215(4539):1501–1503.

Rudwick, M.J.S. *Georges Cuvier, Fossil Bones, and Geological Catastrophes: New Translations and Interpretations of the Primary Texts*. Chicago: University of Chicago Press, 1997.

Rudwick, M.J.S. *The Meaning of Fossils: Episodes in the History of Paleontology*. Chicago: University of Chicago Press, 1976.

Ward, P. (1988). Maastrichtian ammonite and inoceramid ranges from Bay of Biscay Cretaceous-Tertiary boundary sections. In M. Lamolda, M.E. Kauffman, and O. Wallisee, eds., *Paleontology and Evolution: Extinction Events*. Madrid. *Revista Española de Palaontologie*, n. Extraordinario, pp. 119–126.

Ward, P. (1990). The Cretaceous-Tertiary extinctions in the marine realm: a 1990 perspective. In V.L. Sharpton and P.D. Ward, *Global Catastrophes in Earth History*. Boulder, Colorado: *Geological Society of America*. Special Paper 247:425–432.

Ward, P. (1990). A review of Maastrichtian ammonite ranges. Geological Society of America, Special Paper 247:519–530.

Ward, P. (1995). The K/T trial. *Paleobiology* 21(3):245–248.

Ward, P. (1996). After the fall: lessons from the K/T debate. *Palaios* 10:530–538.

Ward, P. (1996). A review of ammonite extinction. In N. Landman and K. Tanabe, eds. *Ammonoid Paleobiology*. New York: Plenum Press, pp. 815–824.

Ward, P. (1997). Impacts and mass extinctions. *Proceedings of the Lawrence Livermore Laboratory*, Special volume, Planetary Defense Workshop, 51–57.

Ward, P., and W. Kennedy (1993). Maastrichtian ammonites from the Biscay region (France and Spain). *Journal of Paleontology* Memoir 34, 67:58.

Ward, P., W.J. Kennedy, K. MacLeod, and J. Mount (1991). Ammonite and *Inoceramid* bivalve extinction patterns in Cretaceous-Tertiary boundary sections of the Biscay Region (southwest France, northern Spain). *Geology* 19:1181–1184.

Ward, P., and K. MacLeod (1988). Macrofossil extinction patterns at Bay of

Biscay Cretaceous-Tertiary boundary sections. Lunar and Planetary Institute Publications, no. 673, 206–207.

Ward, P., J. Wiedmann, and J. Mount (1986). Maastrichtian molluscan biostratigraphy and extinction patterns in a Cretaceous-Tertiary boundary section exposed at Zumaya, Spain. *Geology* 14:899–903.

CHAPTER 2: THE OVERLOOKED EXTINCTION

Aubry, M.-P. (1998). Stratigraphic (dis)continuity and temporal resolution of geological events in the upper Paleocene-lower Eocene deep sea record. In M.-P. Aubry, S. Lucas, and W.A. Berggren, eds., *Late Paleocene–Early Eocene Climatic and Biotic Events in the Marine and Terrestrial Records*. New York: Columbia University Press, pp. 37–66.

Bralower, T.J., J.C. Zachos, E. Thomas, M.N. Parrow, C.K. Paull, D.C. Kelly, I. Premoli-Silva, and W.V. Sliter (1995). Late Paleocene to Eocene paleoceanography of the equatorial Pacific Ocean: stable isotopes recorded at ODP site 865, Allison Guyot. *Paleoceanography* 10:841–865.

Corfield, R.M. (1998). The oxygen and carbon isotopic context of the Paleocene/Eocene epoch boundary. In M.-P. Aubry, S.G. Lucas, and W.A. Berggren, eds., *Late Paleocene–Early Eocene Climatic and Biotic Events in the Marine and Terrestrial Records*, pp. 124–137, Inst. des Sci. de l'Evol., Univ. de Montpellier II, Montpellier, France.

Dickens, G.R., J.R. O'Neil, D.K. Rea, and R.M. Owen (1995). Dissociation of oceanic methane hydrate as a cause of the carbon excursion at the end of the Paleocene. *Paleoceanography* 10:841–865.

Dockery III, D.T. (1998). Molluscan faunas across the Paleocene/Eocene series boundary in the North American Gulf coastal plain. In M.-P. Aubry, S. Lucas, and W.A. Berggren, eds., *Late Paleocene–Early Eocene Climatic and Biotic Events in the Marine and Terrestrial Records*. New York: Columbia University Press, pp. 296–322.

Dockery III, D.T., and P. Lozuet (2003). Molluscan faunas across the Eocene/Oligocene boundary in the North American Gulf coastal plain, with comparison to those of the Eocene and Oligocene of France. In D. Prothero, L. Ivany, E. Nesbitt, eds., *From Greenhouse to Icehouse: The Marine Eocene–Oligocene Transition*. New York: Columbia University Press, pp. 303–340.

Gibson, T.G., and L.M. Bybell (1994). Sedimentary patterns across the Paleocene-Eocene boundary in the Atlantic and Gulf coastal plains of the United States. *Bulletin de la Société Belge de Geologie* 103:237–265.

Hallam, A., and P.B. Wignall (1997). *Mass Extinctions and Their Aftermath*. Oxford: Oxford University Press, p. 320.

Kennett, J.P., and L.D. Stott (1991). Abrupt deep-sea warming, paleoceanographic changes, and benthic extinctions at the end of the Palaeocene. *Nature* 353:319–322.

Kaiho, K. (1991). Global changes of Paleogene aerobic/anaerobic benthic foraminifera and deep-sea circulation. *Palaeogeography, Palaeoclimatology, Palaeoecology* 83:65–85.

Koch, P.L., J.C. Zachos, and P.D. Gingerich (1992). Coupled isotopic changes in marine and continental carbon reservoirs at the Paleocene-Eocene boundary. *Nature* 358:319–322.

Raup, D. (1990). Impact as a general cause of extinction: a feasibility test. In V. Sharpton and P. Ward, eds., *Global Catastrophes in Earth History*. Boulder, Colorado: *Geological Society of America*. Special Paper 247:27–32.

Raup, D. (1991). *Extinction: Bad Genes or Bad Luck?* New York: W.W. Norton.

Raup, D. (1991). A kill curve for Phanerozoic marine species. *Paleobiology* 17:37–48.

Raup, D.M. (1992). Large-body impact and extinction in the Phanerozoic. *Paleobiology* 18(1):80–88.

Raup, D.M., and J.J. Sepkoski (1982). Mass extinctions in the marine fossil record. *Science* 215(4539):1501–1503.

Steineck, P.L., and E. Thomas (1996). The latest Paleocene crisis in the deep-sea: ostracode succession at Maud Rise, Southern Ocean. *Geology* 24:583–586.

Thomas, E. (1998). Biogeography of the Late Paleocene benthic foraminiferal extinction. In M.-P. Aubry, S. Lucas, and W.A. Berggren, eds., *Late Paleocene-Early Eocene Climatic and Biotic Events in the Marine and Terrestrial Records*. New York: Columbia University Press, pp. 214–243.

Thomas, E. (2006). The biogeography of the late Paleocene benthic foraminiferal extinction, In M.-P. Aubry, S. Lucas, and W.A. Berggren, eds., *Late Paleocene–Early Eocene Biotic and Climatic Events in the Marine and Terrestrial Records*. New York: Columbia University Press.

Thomas, E., and N.J. Shackleton (1996). The Palaeocene-Eocene benthic foraminiferal extinction and stable isotope anomalies. *Geological Society London* 101(special publication):401–441.

Zachos, J.C., K.C. Lohmann, J.C.G. Walker, and S.W. Wise (1993). Abrupt climate change and transient climates in the Paleogene: a marine perspective. *Journal of Geology* 100:191–213.

CHAPTER 3: THE MOTHER OF ALL EXTINCTIONS

Baud, A., S. Cirilli, and J. Marcoux (1997). Biotic response to mass extinction: the lowermost Triassic microbialites. *Facies* 36:238–242.

Baud, A., Magaritz, M., and Holser, W.T. (1989) Permian-Triassic of the Tethys: carbon isotope studies. *Geologische Rundschau* 78:649.

Bambach, R.K., A.H. Knoll, and S.C. Wang (2004). Origination, extinction, and mass depletions of marine diversity. *Paleobiology* 30(4):522–542.

Basu, A.R., M.I. Petaev, R.J. Poreda, S.B. Jacobson, and L. Becker (2003). Evidence for an end-Permian catastrophic bolide collision with the earth. *Science* 302:1388–1392.

Becker, L., R.J. Poreda, A.R. Basu, K.O. Pope, T.M. Harrison, C. Nicholson, R. Iasky (2004). Bedout: A possible end-Permian impact crater offshore of northwestern Australia. *Science* 304:1469–1476.

Becker, L., R.J. Poreda, A.G. Hunt, T.E. Bunch, M. Rampino, M. (2001). Impact event at the Permian-Triassic boundary, evidence from extraterrestrial noble gases in fullerenes. *Science* 291:1530–1533.

Benton, M.J., and R.J. Twitchett (2003). How to kill (almost) all life: the end-Permian extinction event. *Trends in Ecology and Evolution* 18:358–365.

Berner, R.A. (2002). Examination of hypotheses for the Permo-Triassic boundary extinction by carbon cycle modeling. *Proceedings of the National Academy of Sciences* 99:4172–4177.

Erwin, D. (1993). *The Great Paleozoic Crisis: Life and Death in the Permian*. New York: Columbia University Press.

Erwin, D. (1994). The Permo-Triassic extinction: *Nature* 367:231–236.

Erwin, D.H. (2002). End-Permian mass extinctions: a review. In C. Koeberl and K.G. MacLeod, eds. *Catastrophic Events and Mass Extinctions: Impacts and Beyond.* Boulder, Colorado: Geological Society of America, Special Paper 356.

Farley, K.A., and S. Mukhopadhyay (2001). An extraterrestrial impact at the Permian-Triassic boundary? Science 293:U1–U3 (comment, online journal edition only).

Farley, K.A., P. Ward, G. Garrison, and S. Mukhopadhyay (2005). Absence of extraterrestrial ^{3}He in Permian-Triassic age sedimentary rocks. *Earth and Planetary Science Letters* 240:265–275.

Gorder, P. 2005. Ohio State University press release: Big bang in Antarctica— killer crater found under ice. http://researchnews.osu.edu/archive/erthboom.htm.

Grice, K., C.Q. Cao, G.D. Love, M.E. Bottcher, R.J. Twitchett, E. Grosjean, R.E. Summons, S.C. Turgeon, W. Dunning, and Y.G. Jin (2005). Photic zone euxinia during the Permian-Triassic superanoxic event. *Science* 307(5710):706–709.

Hallam, A., and P.B. Wignall (1997). *Mass Extinctions and Their Aftermath.* Oxford: Oxford University Press, p. 320.

Huey, R.B., and P.D. Ward (2005). Hypoxia, global warming, and terrestrial late Permian extinctions. *Science* 308(5720):398–401.

Jin, Y.G., Y. Wang, W. Wang, Q.H. Shang, C.Q. Cao, and D.H. Erwin (2000). Pattern of marine mass extinction near the Permian-Triassic boundary in south China. *Science* 289:432–436.

Knoll, A., R. Bambach, D. Canfield, and J. Grotzinger (1996). Comparative earth history and Late Permian mass extinction. *Science* 273:452–457.

Koeberl, C., K.A. Farley, B. Peucker-Ehrinbrink, and M. Sephton (2004). Geochemistry of the Permian-Triassic boundary in Austria and Italy: no evidence for extraterrestrial impact. *Geology* 32:1053–1056.

MacLeod, K.G., R.M.H. Smith, P.L. Koch, and P.D. Ward (2000). Timing of mammal-like reptile extinctions across the Permian-Triassic boundary in South Africa. *Geology* 28:227–230.

Payne, J.L., D.J. Lehrmann, J. Wei, M.J. Orchard, D.P. Schrag, and A.H. Knoll (2004). Large perturbations of the carbon cycle during recovery from the end-Permian extinction. *Science* 305: 506–509.

Renne, P.R., H.J. Melosh, K.A. Farley, W.U. Reimold, C. Koeberl, M.R. Rampino, S.P. Kelly, B.A. Ivanov (2004). Is Bedout an impact crater? Take 2. *Science* 306:610–611.

Retallack, G.J., R.M.H. Smith, and P.D. Ward (2003). Vertebrate extinction across the Permian-Triassic boundary in Karoo Basin, South Africa: *Geological Society of America Bulletin* 115:1133–1152.

Stanley, S.M., and X. Yang (1994). A double mass extinction at the end of the Paleozoic era. *Science* 266:1340–1344.

Steiner, M.B., Y. Eshet, M.R. Rampino, and D.M. Schwindt (2003). Fungal abundance spike and the Permian-Triassic boundary in the Karoo Supergroup (South Africa). *Palaeogeography, Palaeoclimatology, Palaeoecology* 194:405–414.

Summons, R.E., L.L. Jahnke, and Z. Roksandic (1994). Carbon isotopic fractionation in lipids from methanotrophic bacteria—relevance for interpretation of the geochemical record of biomarkers. *Geochimica Et Cosmochimica Acta* 58(13):2853–2863.

Ward, P., Montgomery, D., and Smith, R. (2000). Altered river morphology in South Africa related to the Permian-Triassic extinction. *Science* 289:1740–1743.

Ward, P.D., J. Botha, R. Buick, M.O. De Kock, D.H. Erwin, G.H. Garrison, J.L. Kirschvink, and R. Smith. 2005. Abrupt and gradual extinction among late Permian land vertebrates in the Karoo Basin, South Africa. *Science* 307(5710):709–714.

White, R.V. (2002). Earth's biggest 'whodunnit': unravelling the clues in the case of the end-Permian mass extinction: *Philosophical Transactions of the Royal Society of London, Series B* 360:2963–2985.

CHAPTER 4: THE MISINTERPRETED EXTINCTION

Galli, M.T., F. Jadoul, S.M. Bernasconi, and H. Weissert (2005). Anomalies in global carbon cycling and extinction at the Triassic/Jurassic boundary: evidence from a marine C-isotope record. *Palaeogeography Palaeoclimatology Palaeoecology* 216(3–4):203–214.

Hallam, A. (1981). The End-Triassic Bivalve Extinction Event. *Palaeogeography Palaeoclimatology Palaeoecology* 35(1):1–44.

Hallam, A. (2002). How catastrophic was the end-Triassic mass extinction? *Lethaia* 35(2):147–157.

Hallam, A., and P.B. Wignall (1997). *Mass Extinctions and Their Aftermath.* Oxford: Oxford University Press, p. 320.

Hallam, A., and P.B. Wignall (2000). Facies changes across the Triassic-Jurassic boundary in Nevada, USA. *Journal of the Geological Society* 157:49–54.

Hesselbo, S.P., S.A. Robinson, F. Surlyk, and S. Piasecki (2002). Terrestrial and marine extinction at the Triassic-Jurassic boundary synchronized with major carbon-cycle perturbation: a link to initiation of massive volcanism? *Geology* 30(3):251–254.

Hodych, J.P., and G.R. Dunning (1992). Did the Manicouagan impact trigger end-of-Triassic mass extinction? *Geology* 20(1):51–54.

Lucas, S.G. (1998). Global Triassic tetrapod biostratigraphy and biochronology. *Palaeogeography, Palaeoclimatology, Palaeoecology* 143:347–384.

Marzoli, A., P.R. Renne, E.M. Piccirillo, M. Ernesto, G. Bellieni, and A. De Min (1999). Extensive 200-million-year-old continental flood basalts of the Central Atlantic Magmatic Province. *Science* 284(5414):616–618.

McElwain, J.C., D.J. Beerling, and F.I. Woodward (1999). Fossil plants and global warming at the Triassic-Jurassic boundary. *Science* 285(5432):1386–1390.

Morgan, J.T., J. Reston, C.R. Ranero (2004). Contemporaneous mass extinctions, continental flood basalts, and 'impact signals': are mantle plume-induced lithospheric gas explosions the causal link? *Earth and Planetary Science Letters* 217:263–284.

Olsen, P.E., and D.V. Kent (1996). Milankovitch climate forcing in the tropics of Pangaea during the Late Triassic. *Palaeogeography Palaeoclimatology Palaeoecology* 122(1–4):1–26.

Olsen, P.E., D.V. Kent, H.D. Sues, C. Koeberl, H. Huber, A. Montanari, E.C. Rainforth, S.J. Fowell, M.J. Szajna, and B.W. Hartline (2002). Ascent of dinosaurs linked to an iridium anomaly at the Triassic-Jurassic boundary. *Science* 296(5571):1305–1307.

Palfy, J., A. Demeny, J. Haas, M. Hetenyi, M.J. Orchard, and I. Veto (2001). Carbon isotope anomaly and other geochemical changes at the Triassic-Jurassic boundary from a marine section in Hungary. *Geology* 29(11):1047–1050.

Palfy, J., J.K. Mortensen, E.S. Carter, P.L. Smith, R.M. Friedman, and H.W. Tipper (2000). Timing the end-Triassic mass extinction: first on land, then in the sea? *Geology* 28(1):39–42.

Payne, J.L., D.J. Lehrmann, J. Wei, M.J. Orchard, D.P. Schrag, and A.H. Knoll (2004). Large perturbations of the carbon cycle during recovery from the end-Permian extinction: *Science* 305:506–509.

Tanner, L.H., S.G. Lucas, and M.G. Chapman. 2004. Assessing the record and causes of late Triassic extinctions. *Earth-Science Reviews* 65(1–2):103–139.

Ward, P.D., G.H. Garrison, J.W. Haggart, D.A. Kring, and M.J. Beattie (2004). Isotopic evidence bearing on late Triassic extinction events, Queen Charlotte Islands, British Columbia, and implications for the duration and cause of the Triassic/Jurassic mass extinction. *Earth and Planetary Science Letters* 224(3–4):589–600.

Ward, P.D., J.W. Haggart, E.S. Carter, D. Wilbur, H.W. Tipper, and T. Evans (2001). Sudden productivity collapse associated with the Triassic-Jurassic boundary mass extinction. *Science* 292(5519):1148–1151.

CHAPTER 5: A NEW PARADIGM FOR MASS EXTINCTION

Bambach, R.K., A.H. Knoll, and S.C. Wang (2004). Origination, extinction, and mass depletions of marine diversity. *Paleobiology* 30(4):522–542.

Beauchamp, B., and A. Baud (2002). Growth and demise of Permian biogenic chert along northwest Pangea: evidence for end-Permian collapse of thermohaline circulation. *Palaeogeography, Palaeoclimatology, Palaeoecology* 184:37–63.

Berner, R.A., and D.E. Canfield (1989). A model for atmospheric oxygen over Phanerozoic time. *American Journal of Science* 289:333–361.

Broecker, W.S. (2003). Does the trigger for abrupt climate change reside in the ocean or in the atmosphere? *Science* 300:1519–1522.

Canfield, D.E., and R.A. Berner (1987). Dissolution and pyritization of magnetite in anoxic marine sediments: *Geochimica et Cosmochimica Acta* 51:645–659.

Isozaki, Y. (1997). Permo-triassic boundary superanoxia and stratified superocean: records from lost deep sea. *Science* 276(5310):235–238.

Kidder, D.L., and T.R. Worsley (2003). Late Permian warming, the rapid latest Permian transgression, and the Permo-Triassic extinction. Geological Society of America Annual Meeting: Seattle, Washington.

Kiehl, J., and C. Shields (2005). Climate simulation of the latest Permian: implications for mass extinction. *Geology* 9:757–760.

Knoll, A., R. Bambach, D. Canfield, and J. Grotzinger (1996). Comparative earth history and late Permian mass extinction. *Science* 273:452–457.

Kump L.R., A. Pavlov, M.A. Arthur (2005). Massive release of hydrogen sulfide to the surface ocean and atmosphere during intervals of oceanic anoxia. *Geology* 33(5):397–400.

Looy, C., W.A. Brugman, D.L. Dilcher, and H. Visscher (1999). The delayed resurgence of equatorial forests after the Permian-Triassic ecologic crisis. *Proceedings of the National Academy of Sciences* 96:13857–13862.

Looy, C., R.J. Twitchett, D.L. Dilcher, J.H.A. Van Konijnenburg-Van Cittert, and H. Visscher (2001). Life in the end-Permian dead zone. *Proceedings of the National Academy of Sciences* 98:7879–7883.

Marzoli, A., P.R. Renne, E.M. Piccirillo, M. Ernesto, G. Bellieni, and A. De Min (1999). Extensive 200-million-year-old continental flood basalts of the Central Atlantic Magmatic Province. *Science* 284(5414):616–618.

McElwain, J.C., D.J. Beerling, and F.I. Woodward (1999). Fossil plants and global warming at the Triassic-Jurassic boundary. *Science* 285(5432):1386–1390.

Schmidt, M.W., H.J. Spero, and D.W. Lea (2004). Links between salinity variation in the Caribbean and North Atlantic thermohaline circulation, *Nature* 428:160–163.

Summons, R.E., L.L. Jahnke, and Z. Roksandic (1994). Carbon isotopic fractionation in lipids from methanotrophic bacteria—relevance for interpretation of the geochemical record of biomarkers. *Geochimica et Cosmochimica Acta* 58(13):2853–2863.

Talley, L.D. (2003). Shallow, intermediate and deep overturning components of the global heat budget, *Journal of Physical Oceanography* 33:530–560.

Thiel, V., J. Peckmann, H.H. Richnow, U. Luth, J. Reitner, and W. Michaelis (2001). Molecular signals for anaerobic methane oxidation in Black Sea seep carbonates and a microbial mat. *Marine Chemistry* 73(2):97–112.

Vellinga, M., and R.A. Wood (2002). Global climatic impacts of a collapse of the Atlantic thermohaline circulation. *Climatic Change* 54:251–267.

Vellinga, M., R.A. Wood, and J.M. Gregory (2002). Processes governing the recovery of a perturbed thermohaline circulation in HadCM3. *Journal of Climate* 15:764–780.

Ward, P. (2006). Impact from the deep. *Scientific American* October:64–71.

White, R.V., and A.D. Saunders (2005). Volcanism, impact and mass extinctions: incredible or credible coincidences? *Lithos* 79(3–4):299–316.

Winguth, A.M.E., C. Heinze, J.E. Kutzbach, E. Maier-Reimer, and U. Mikolajewicz (2002). Simulated warm polar currents during the middle Permian. *Paleoceanography* 17:9–1, 9–15.

Xie, S.C., R.D. Pancost, H.F. Yin, H.M. Wang, and R.P. Evershed (2005). Two episodes of microbial change coupled with Permo/Triassic faunal mass extinction. *Nature* 434(7032):494–497.

CHAPTER 6: THE DRIVER OF EXTINCTION

Berner, R.A. (2003). The long-term carbon cycle, fossil fuels and atmospheric composition. *Nature* 426:323–326.

Berner, R.A. The Phanerozoic Carbon Cycle. New York: Oxford University Press.

Berner, R.A., and Z. Kothavala (2001) Geocarb III: a revised model of atmospheric CO_2 over Phanerozoic time. *American Journal of Science* 301:182–204.

Berner, R.A., and Raiswell, R.A. (1983). Burial of organic-carbon and pyrite sulfur over Phanerozoic time—a new theory: *Geochimica et Cosmochimica Acta* 47:855–862.

Brocks J.J., and Summons R.E. (2003). Sedimentary hydrocarbons, biomarkers for early life. In H.D. Holland, ed., *Treatise in Geochemistry*. Copenhagen: Elsevier, vol. 8, p. 53.

Cerling, T.E. (1992). Use of carbon isotopes in paleosols as an indicator of the $P(CO_2)$ of the paleoatmosphere. *Global Biogeochemical Cycles* 6:307–314.

Coccioni, R., and S. Galeotti (2003). Deep-water benthic foraminiferal events from the Massignano Eocene/Oligocene boundary stratotype, central Italy. In D. Prothero, L. Ivany, E. Nesbitt, eds., *From Greenhouse to Icehouse: The Marine Eocene–Oligocene Transition*. New York: Columbia University Press, pp. 438–452.

Diester-Haas, L., and J. Zachos (2003). The Eocene-Oligocene transition in the equatorial Atlantic (ODP site 925): paleoproductivity increase and positive $\sigma^{13}C$ excursion. In D. Prothero, L. Ivany, E. Nesbitt, eds., *From Greenhouse to Icehouse: The Marine Eocene–Oligocene Transition*. New York: Columbia University Press, pp. 397–418.

Eagan, B. (2004) The Long Summer. New York: Basic Books.

Hinrichs, K.U., L.R. Hmelo, and S.P. Sylva (2003). Molecular fossil record of elevated methane levels in late Pleistocene coastal waters. *Science* 299(5610):1214–1217.

McManus, J.F., R. Francois, R., J.-M. Gherardi, L.D. Kelgwin, and S. Brown-Leger, S. (2004). Collapse and rapid resumption of Atlantic meridional circulation linked to deglacial climate changes. 428:834–837.

Rind, D., G. Russell, G. Schmidt, S. Sheth, D. Collins, P. deMenocal, and J. Teller (2001). Effects of glacial meltwater in the GISS coupled atmosphere-ocean model 1, North Atlantic Deep Water response. *Journal of Geophysical Research* 106:27335–27353.

Schneider, S.H., and R.S. Chen (1980). Carbon dioxide flooding: physical factors and climatic impact. *Annual Review of Energy* 5:107–140.

Seager, R., and D.S. Battisti (2006). Challenges to our understanding of the general circulation: abrupt climate change. In Schneider and Sobel, eds., *Global Circulation of the Atmosphere*. Princeton, New Jersey: Princeton University Press.

Ward, P., and Brownlee, D. (2003). *The Life and Death of Planet Earth*. New York: Henry Holt.

CHAPTER 7: BRIDGING THE DEEP PAST AND NEAR PAST

Blunier, T., and E.J. Brook (2001). Timing of millennial-scale climate change in Antarctica and Greenland during the last glacial period. *Science* 291:109–112.

Boccaletti, G., R. Ferreira, A. Adcroft, D. Ferreira, and J. Marshall (2005). The vertical structure of ocean heat transport, *Geophysical Research Letters* 32(10):L10603.1–L10603.4.

Boccaletti, G., R.C. Pacanowski, S.G.H. Philander, and A.V. Fedorov (2004). The thermal structure of the upper ocean, *Journal of Physical Oceanography* 34:888–902.

Bonani, G. (1993). Correlations between climate records from North Atlantic sediments and Greenland ice. *Nature* 365:143–147.

Bond, G., W. Broecker, S. Johnsen, J. McManus, L. Labeyrie, J. Jouzel, and G. Bonani (1993). Correlations between climate records from North Atlantic sediments and Greenland ice. *Nature* 365:143–147.

Bond, G., W. Showers, M. Cheseby, R. Lotti, P. Almasi, P. deMenocal, P. Priore, H. Cullen, I. Hajdas, and G. Bonani (1997). A pervasive millennial-scale cycle in North Atlantic Holocene and glacial climates. *Science* 278:1257–1266.

Broecker, W.S. (2003). Does the trigger for abrupt climate change reside in the ocean or in the atmosphere? *Science* 300:1519–1522.

Broecker, W.S., and G.H. Denton (1990). The role of ocean-atmosphere reorganization in glacial cycles. *Quaternary Science Reviews* 9:305–341.

Broecker, W.S., D.M. Peteet, and D. Rind (1985). Does the ocean-atmosphere system have more than one stable mode of operation? *Nature* 315:21–26.

Brook, E.J., S. Harder, J. Severinghaus, and M. Bender (1999). Atmospheric methane and millennial-scale climate change. In P.U. Clark, R.S. Webb, and L.D. Keigwin, eds., *Mechanisms of Global Climate Change at Millennial Time Scales*. Washington, D.C.: American Geophysical Union, pp. 165–175.

Cane, M.A. (1984). Modeling sea level during El Niño. *Journal of Physical Oceanography* 14:1864–1874.

Clark, P.U., S.J. Marshall, G.K.C. Clarke, S.W. Hostetler, J.M. Licciardi, J.T. and Teller (2001). Freshwater forcing of abrupt climate change during the last glaciation. *Science* 293:283–287.

Clarke, G.K.C., S.J. Marshall, C. Hillaire-Marcel, G. Bilodeau, C. and Veiga-Pires (1999). A glaciological perspective on Heinrich events. In P.U. Clark, R.S. Webb, and L.D. Keigwin, eds., *Mechanisms of Global Climate Change at Millennial Time Scales*. Washington D.C.: American Geophysical Union, pp. 243–262.

Cuffey, K.M., G.D. Clow, R.B. Alley, M. Stuiver, E.D. Waddington, and R.W. Saltus (1995). Large arctic temperature change at the Wisconsin-Holocene glacial transition. *Nature* 270:455–458.

Denton, G.H., R.B. Alley, G.C. Comer, and W.S. Broecker (2005). The role of seasonality in abrupt climate change. *Quaternary Science Reviews* 24:1159–1182.

Driscoll, N.W., and G.H. Haug (1998). A short circuit in thermohaline circulation: a cause for Northern Hemisphere glaciation? *Science* 282:436–438.

Ganapolski, A., and S. Rahmstorff (2001). Rapid changes of glacial climate simulated in a coupled climate model, *Nature* 409:153–158.

Gordon, A.L., R.F. Weiss, W.M. Smethie, and M.J. Warner (1992). Thermocline and intermediate water communication between the South Atlantic and Indian Oceans. *Journal of Geophysical Research* 97:7223–7240.

Hall, J., B. Dong, and P.J. Valdes (1996). Atmospheric equilibrium, instability and energy transport at the last glacial maximum. *Climate Dynamics* 12:497–511.

Hays J.D., J. Imbrie, and N.J. Shackleton (1976). Variations in the Earth's orbit, pacemaker of the ice ages. *Science* 194:1121–1132.

Hemming, S.R. (2004). Heinrich events: massive late Pleistocene detritus layers of the North Atlantic and their global imprint. *Reviews of Geophysics* 42:1–42.

Hughen, K.A., J.T. Overpeck, S.J. Lehman, M. Kashgarian, J. Southon, L.C. Peterson, R. Alley, and D.M. Sigman (1998). Deglacial changes in ocean circulation from an extended radiocarbon calibration. *Nature* 391:65–68.

Hughen, K.A., J.T. Overpeck, L.C. Peterson, and S. Trumbore (1996). Rapid climate changes in the tropical Atlantic region during the last deglaciation. *Nature* 380:51–54.

Hughen, K.A., J. Southon, S.J. Lehman, and J.T. Overpeck (2000). Synchronous radiocarbon and climate shifts during the last deglaciation. *Science* 290:1951–1954.

Keeling, C.D., T.P. Whorf, M. Wahlen, and J van der Plicht. 1995. Interannual extremes in the rate of atmospheric carbon dioxide since 1980. *Nature* 375:666–670.

Kim, J.-H., and Schneider, R.R. (2003). Low-latitude control of interhemispheric sea-surface temperature contract in the tropical Atlantic over the past 21 k years: the possible role of SE trade winds. *Climate Dynamics* 21:337–347.

Lea, D.W., D.P. Pak, L.C. Peterson, and K.A. Hughen (2003). Synchroneity of tropical and high-latitude Atlantic temperatures over the last glacial termination. *Science* 301:1361–1364.

McManus, J.F., R. Francois, J.-M. Gherardi, L.D. Kelgwin, and S. Brown-Leger (2004). Collapse and rapid resumption of Atlantic meridional circulation linked to deglacial climate changes. *Nature* 428:834–837.

Mercer, J.H. (1978). West Antarctic ice sheet and CO_2 greenhouse effect: a threat of disaster? *Nature* 271:321–325.

Peterson, L.C., G.H. Haug, K.A. Hughen, and U. Rohl (2000). Rapid changes in the hydrologic cycle of the tropical Atlantic during the last glacial. *Science* 290:1947–1951.

Renssen, H., and R.F.B. Isarin (2001). The two major warming phases of the last deglaciation at 14.7 and 11.5 kyr cal BP in Europe: climate reconstructions and AGCM experiments. *Global and Planetary Change* 30:117–154.

Ruhlemann, C., S. Mulitza, P.J. Müller, G. Wefer, and R. Zhan (1999). Warming of the tropical Atlantic Ocean and slowdown of thermohaline circulation during the last deglaciation. *Nature* 402:511–514.

Sachs, J.P., and S.J. Lehman (1999). Subtropical North Atlantic temperatures 60,000 to 30,000 years ago *Science* 286:756–759.

Schmidt, M.W., H.J. Spero, and D.W. Lea (2004). Links between salinity variation in the Caribbean and North Atlantic thermohaline circulation. *Nature.* 428:160–163.

Seager, R., D.S. Battisti (2005). Challenges to our understanding of the general circulation: abrupt climate change. In Schneider and Sobel, eds., *Global Circulation of the Atmosphere*. Princeton, New Jersey: Princeton University Press.

Seager, R., D.S. Battisti, J. Yin, N. Gordon, N.H. Naik, A.C. Clement, and M.A. Cane (2002). Is the Gulf Stream responsible for Europe's mild winters? *Quarterly Journal of the Royal Meteorological Society* 128:2563–2586.

Talley, L.D. (2003). Shallow, intermediate and deep overturning components of the global heat budget, *Journal of Physical Oceanography* 33:530–560.

Timmermann, A., S.-I. An, U. Krebs, and H. Goosse (2005). ENSO suppression due to weakening of the North Atlantic thermohaline circulation. *J. Climate*, 18, 3122–3139.

Vellinga, M., and R.A. Wood, (2002). Global climatic impacts of a collapse of the Atlantic thermohaline circulation. *Climatic Change* 54:251–267.

Vellinga, M., R.A. Wood, and J.M. Gregory (2002). Processes governing the recovery of a perturbed thermohaline circulation in HadCM3, *Journal of Climate* 15:764–780.

Warren, B.A. (1983). Why is no deep water formed in the North Pacific? *Journal of Marine Research* 41:21–26.

Winton, M. (2003). On the climatic impact of ocean circulation. *Journal of Climate* 16:2875–2889.

Wright, J.D., and K.G. Miller (1996). Control of North Atlantic deep water circulation by the Greenland-Scotland Ridge. *Paleoceanography* 11:157–170.

Wright, J.D., K.G. Miller, R.G. Fairbanks (1991). Evolution of modern deep-water circulation: evidence from the late Miocene Southern Ocean. *Paleoceanography* 6:275–290.

Wunsch, C. (2003). Greenland-Antarctic phase relations and millennial time-scale climate fluctuations in the Greenland ice-cores. *Quaternary Science Reviews* 22:1631–1646.

CHAPTER 8: THE ONCOMING EXTINCTION OF WINTER

Keeling, C.D., T.P. Whorf, M. Wahlen, and J. van der Plicht (1995). Interannual extremes in the rate of atmospheric carbon dioxide since 1980. *Nature* 375:666–670.

Roe, G.H., and Steig, E.J. (2004). Characterization of millennial-scale climate variability. *Journal of Climate* 17:1929–1944.

Ruddimen, W.F. (2005). *Plows, Plagues, and Petroleum*. Princeton, N.J.: Princeton University Press.

CHAPTER 9: BACK TO THE EOCENE

Alley, R. (2005). *The Two Mile Time Machine*. Princeton, N.J.: Princeton University Press.

Barth, M.C., and J.G. Titus (eds.) (1984). *Greenhouse Effect and Sea Level Rise: A Challenge for This Generation*. New York: Van Nostrand Reinhold.

Bates, T. (1999). Sink or Swim. *Asbury Park Press*, Millennium Section at 1 (July 4, 1999). Also available at http://www.injersey.com/2000/story/1,2297,195727,00.html.

El-Raey, M., S. Nasr, O. Frihy, S. Desouki, and Kh. Dewidar (1995). Potential impacts of accelerated sea-level rise on Alexandria Governorate, Egypt. *Journal of Coastal Research* 14(special issue):190–204.

Environmental Protection Agency (1983). Projecting future sea level rise. Washington, D.C.: Strategic Studies Staff, U.S. Environmental Protection Agency.

Environmental Protection Agency (1989). The potential impacts of global climate change on the United States. Washington, D.C.: United States Environmental Protection Agency.

Environmental Protection Agency (1995). The probability of sea level rise. Washington, D.C.: Environmental Protection Agency. Also available at http://www.epa.gov/globalwarming/publications/impacts/sealevel/probability.html.

Etheridge, D.M., L.P. Steele, R.L. Langenfelds, R.J. Francey, J.-M. Barnola, and V.I. Morgan. 1996. Natural and anthropogenic changes in atmospheric CO_2 over the last 1000 years from air in Antarctic ice and firn. *Journal of Geophysical Research* 101:4115–4128.

Federal Emergency Management Agency, Federal Insurance Administration (1991). Projected impact of relative sea level rise on the National Flood Insurance Program. Washington, D.C.: Federal Emergency Management Agency.

Han, M., J. Hou, and L. Wu (1995). Potential impacts of sea level rise on China's coastal environment and cities: a national assessment. *Journal of Coastal Research* 14(special issue):79–95.

Hoerling, M.P., J.S. Whitaker, A. Kumar, W. Wang (2004). Twentieth century climate change. Part II: Understanding the effect of Indian Ocean warming. *Climate Dynamics* 23:391–405.

Hoskins, B.J., and P.J. Valdes (1990). On the existence of storm tracks, *Journal of the Atmospheric Sciences* 47:1854–1864.

Huq, A, S.I. Ali, and A.A. Rahman (1995). Sea level rise and Bangladesh: a preliminary analysis. *Journal of Coastal Research* 14(special issue):44–53.

Hurrell, J.W., M.P. Hoerling, A.S. Phillips, and T. Xu (2004). Twentieth century North Atlantic climate change. Part I: assessing determinism. *Climate Dynamics* 23:371–389.

Landsea, C.W. (1999). FAQ : Hurricanes, typhoons, and tropical cyclones. Part G: tropical cyclone climatology. Miami: National Hurricane Center. Also available at http://www.aoml.noaa.gov/hrd/tcfaq/tcfaqG.html#G9.

Intergovernmental Panel on Climate Change (1990). *The IPCC Scientific Assessment.* Cambridge and New York: Cambridge University Press.

Intergovernmental Panel on Climate Change (1996a). *Climate Change 1995: The Science of Climate Change.* Cambridge and New York: Cambridge University Press.

Intergovernmental Panel on Climate Change (1996b). *Climate Change 1995: Impacts, Adaptations, and Mitigation of Climate Change.* Cambridge and New York: Cambridge University Press.

Intergovernmental Panel on Climate Change (1998). *The Regional Effects of Climate Change.* Cambridge and New York: Cambridge University Press.

Mercer, J.H. (1978). West Antarctic ice sheet and CO_2 greenhouse effect: a threat of disaster? *Nature* 271:321–325.

Robinson, A.B., S.L. Baliunas, W. Soon, and Z.W. Robinson (1998). Environmental effects of increased atmospheric carbon dioxide. http://www.oism.org/pproject/s33p36.htm.

Schneider, E.K., L. Bengtsson, and Z.Z. Hu (2003). Forcing of Northern Hemisphere climate trends. *Journal of the Atmospheric Sciences* 60:1504–1521.

Schubert, S.D., M.J. Suarez, P.J. Region, R.D. Koster, and J.T. Bacmeister (2004). Causes of long-term drought in the United States Great Plains. *Journal of Climate* 17:485–503.

Timmermann, A., S.-I. An, U. Krebs, and H. Goosse (2005). ENSO suppression due to weakening of the North Atlantic thermohaline circulation. *Journal of Climate* 18:3122–3139.

Titus, J.G. (1998). Rising seas, coastal erosion, and the takings clause: how to save wetlands and beaches without hurting property owners. *Maryland Law Review* 57:1281–1398. Also available at http://www.epa.gov/globalwarming/publications/impacts/sealevel/index.html.

Titus, J.G., and V. Narayanan (1996). The risk of sea level rise: a delphic Monte Carlo analysis in which twenty researchers specify subjective probability distributions for model coefficients within their respective areas of expertise. *Climatic Change* 33:151–212. Also available at http://www.epa.gov/globalwarming/publications/impacts/sealevel/index.html.

Titus, J.G., R.A. Park, S. Leatherman, R. Weggel, M.S. Greene, M. Treehan, S. Brown, and C. Gaunt, G. Yohe (1991). Greenhouse effect and sea level rise: the cost of holding back the sea. *Coastal Management* 19:3:171–204. Also available at http://www.epa.gov/globalwarming/publications/impacts/sealevel/index.html.

Titus J., C. Richman (2000). Maps of lands vulnerable to sea level rise: modeled elevations along the U.S. Atlantic and Gulf Coasts. *Climate Research* 18:205–228.

Yohe, G. (1990). The cost of not holding back the sea. *Coastal Management* 18:403–432.

Yohe, G., J. Neumann, P. Marshall, and H. Ameden (1996). The economic cost of greenhouse-induced sea-level rise for developed property in the United States. *Climatic Change* 32:387–410.

Index